NANOPHOTONICS

DEVICES, CIRCUITS, AND SYSTEMS

NANOPHOTONICS

Preecha P. Yupapin   Keerayoot Srinuanjan   Surachart Kamoldilok

PAN STANFORD PUBLISHING

Published by

Pan Stanford Publishing Pte. Ltd.
Penthouse Level, Suntec Tower 3
8 Temasek Boulevard
Singapore 038988

Email: editorial@panstanford.com
Web: www.panstanford.com

British Library Cataloguing-in-Publication Data
A catalogue record for this book is available from the British Library.

Nanophotonics: Devices, Circuits, and Systems

ISBN 978-981-4364-36-2 (Hardcover)
ISBN 978-981-4364-37-9 (eBook)

Printed in the USA

Contents

Preface

Behaviors of light in small-scale optics or nano/micro-optical devices have shown promising results which can be useful for many fundamental and applied researches, especially in nanoelectronics. In this book, a new design of a small-scale optical device, a microring resonator, is proposed. Also presented is a design which can be used to generate forms of light on a chip, for applications such as optical spin, antennas, and whispering gallery modes. Most of the chapters use the proposed device made up of silica and InGaAsP/InP with a linear optical add-drop filter incorporating two nonlinear micro/nano rings on both sides of the center ring (add-drop filter). This particular configuration is known as a PANDA ring resonator. A light pulse, for instance Gaussian, bright, and dark solitons, is fed into the system through different ports such as the add port and the through port. By using practical device parameters, the simulation results are obtained using the OptiWave and MATLAB programs. Results obtained by both analytical and numerical methods show that many applications can be exploited. By using practical device parameters, such devices can be fabricated and implemented in the near future. The book shows that interdisciplinary use of the proposed system can be available for many applications, especially when the device is coated with a metallic material.

Preecha Yupapin and Jalil Ali
Spring 2013

Small-Scale Optics

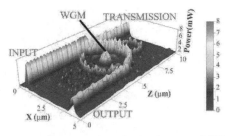

Figure   Whispering gallery modes (WGMs) in a PANDA ring planar waveguide, which is the key concept of this book.

Chapter 1

Nanophotonics

1.1 Nanophotonics

Nanophotonics is the study of the behavior of light on the nano-meter scale, which is considered a branch of optical engineering that deals with optics, or the interaction of light with particles or substances, at deeply subwavelength scales. To increase the resolution of optical measurement methods, the behavior of light in the nanoscale regime was investigated. The scientific and industrial communities are becoming more interested in the characterization of materials and phenomena within the scale of a few nanometers. So, alternative techniques must be utilized. The ability to fabricate devices on the nanoscale, which have been developed recently, provided the catalyst for this area of study.

The field of nanoscale photonics has experienced significant growth in the past decade, which will give a broad introduction to some basic interests of nanophotonic devices. This is a typically critical feature size on the order of ~1–500 nm. The operation of the nanophotonic device is described, in which the high refractive index structure has led to significantly smaller optical waveguides and microscopic optical cavities, and the concept of photonic band gap structures. These nanoscale photonic structures open up various possibilities for novel photonic devices. The use of nano-optical structures enables the control of atomic emissions, which have been used to realize microcavity lasers with very low lasing thresholds and controlled photon statistics. There are many

Nanophotonics: Devices, Circuits, and Systems
Preecha P. Yupapin, Keerayoot Srinuanjan, and Surachart Kamoldilok
Copyright © 2013 Pan Stanford Publishing Pte. Ltd.
ISBN 978-981-4364-36-2 (Hardcover), 978-981-4364-37-9 (eBook)
www.panstanford.com

applications of nanophotonic devices, for example, photonic-wire nanolasers, quantum dots, microdisk lasers, and random lasers. Nanophotonic structures also enable much smaller device sizes and higher integration densities, leading to high-density optoelectronic integrated circuits that may be useful for broadband data interconnects on chip. Device and integration examples include photonic crystal lasers, coupled optical resonator devices, and devices with high nonlinear optical interactions.

Moreover, nowadays, scientists and engineers are interested in the various topics of nanophotonic devices, including subwavelength elements, plasmonic devices, silicon nanophotonic devices and systems, diffractive elements, nanophotonic biosensors, nanocavity devices, photovoltaic cells, nano-opto-electromechanical systems (NOEMS), resonance-based devices, waveguide devices, slow-light elements, nano-antennas, nanostructured lasers, and nanostructured detectors. For instance, the recent development in plasmonic nanophotonics opens up further avenues to engineer the propagation of light beams, including near-field imaging with a negative-refractive-index lens. The potential applications of some of these devices will be discussed and near-future nanophotonic structures described.

1.2 The Basic Concept of Nanophotonics

Consider the interaction between light (or photons) and electrons by the use of a microscopic model. The momentum of an electron is defined by

$$\vec{\mathbf{p}} = \hbar \vec{\mathbf{k}} \tag{1.1}$$

Here, $\vec{\mathbf{k}}$ is the wave vector of the electron state (Plank's constant $h = 6.624 \times 10^{-27}$ ergs and $\hbar = h/2\pi$), whereas the classical momentum of a free electron is mv, where m denotes the mass and v the velocity. When a charge of mass m moves, it creates electric fields and then creates photons, or light. On the other hand, the electric fields make the charges move, so we can detect the light. These discrete quanta, or relatively localized units of optical energy, maintain their identity as units throughout the processes of emission, transmission, reflection, diffraction, absorption, etc. The photon is described in a microscopic model by defining its mass and momentum as follows:

First, we consider a photon of light in a vacuum. The energy E of the photon is given by

$$E = h\upsilon \tag{1.2}$$

Here, υ is the frequency of the radiation in s^{-1}. If Eq. 1.2 is combined with the well-known relation, then

$$c = \upsilon\lambda_0 \tag{1.3}$$

Here, c and λ_0 are the speed and wavelength of light in vacuum, respectively. The result is that the energy of the photon is given by

$$E = \frac{hc}{\lambda_0} \tag{1.4}$$

There are some similarities between photons and electrons. Consider the propagation of photons in dielectrics material or in crystals. The basic equation describing the wavelength of light is

$$\lambda = \frac{h}{p} = \frac{c}{\upsilon} \tag{1.5}$$

And the equation describing the wavelength of electrons is

$$\lambda = \frac{h}{p} = \frac{h}{mv} \tag{1.6}$$

Then, we can compare the equation of motion between photons and electrons that propagate in free space by the use of Maxwell's equations for photons and Schrödinger's equation for electrons.

From Maxwell's equations

$$\nabla \times H = \frac{1}{c}\frac{\partial D}{\partial t} \text{ and } \nabla \times E = \frac{-1}{c}\frac{\partial B}{\partial t} \tag{1.7}$$

where B and H are the magnetic field and magnetic field strength, respectively, D is electric displacement field, and t is the time interval. For a plane wave

$$\nabla \cdot \nabla E(r) = \varepsilon k_0^2 E(r) \tag{1.8}$$

where the solution is in the form of a photon plane wave:

$$E = E_0 \left(e^{ik \cdot r - \omega t} + e^{-ik \cdot r + \omega t} \right) \tag{1.9}$$

And Schrödinger's equation that allows us to calculate the energy of an electron is given by

$$\frac{-\hbar^2}{2m}[\nabla \cdot \nabla + V(r)]\psi(r) = E\psi(r) \tag{1.10}$$

where the solution is in the form of an electron plane wave, given by

$$\psi = c\left(e^{ik \cdot r - \omega t} + e^{-ik \cdot r + \omega t}\right) \tag{1.11}$$

The above equations have solutions in similar forms. Finally, we can conclude that free space propagation of both electrons and photons can be described by plane waves. There momentums are in the same analogues: $\vec{p} = \hbar \vec{k}$, where $k = 2\pi/\lambda$ for a photon and $k = (2\pi/h)mv$ for an electron. The energy is $E = pc$ for both photons and electrons. But, on the other hand, the propagation of light (or photon) is affected by the dielectric medium (refractive index), while the propagation of electrons is affected by Coulomb potential. Thus, we have some similarity in the properties of both photons and electrons.

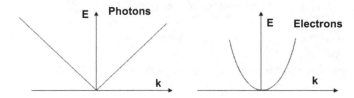

Figure 1.1   Free space dispersion for photons and electrons.

1.3   Near-Field Optics

Near-field optics is the branch of optics that considers configurations that depend on the passage of light near an element with subwavelength features and the coupling of that light to a second element located at a subwavelength distance from the first. The barrier of spatial resolution imposed by the very nature of light itself in conventional optical microscopy contributed significantly to the development of near-field optical devices, most notably the near-field scanning optical microscope (NSOM). The limit of optical resolution in a conventional microscope, the so-called diffraction limit, is on the order of half the wavelength of the light used to image. Thus, when imaging at visible wavelengths, the smallest resolvable objects are several hundred nanometers in size. Using near-field optical techniques, researchers currently resolve features on the order of tens of nanometers in size. While other imaging techniques (e.g., atomic force microscopy and electron microscopy) can resolve features of a much smaller size, the many advantages of optical microscopy make near-field optics a field of considerable interest.

The notion of developing a near-field optical device was first conceived by Synge in 1928 [1] but was not realized experimentally until the 1950s, when several researchers demonstrated the feasibility of subwavelength resolution. The common feature of all NSOMs is the nanometer-sized detector able to collect or emit photons after coupling with a subwavelength-size object deposited on a surface, which depends on the experimental design. This nanodetector can be used to transmit the collected light to an appropriate macrodetector (e.g., a photomultiplier) located far from the object [2]. Today, many experimental configurations based on this concept of nanodetection provide us with an increasing amount of optical information about the nanoworld. A recent comprehensive review of the experimental aspects of near-field microscopy is found in reference [3].

The scanning near-field optical microscope (SNOM) brought a new powerful tool in the field of imaging and surface science. The ability to investigate the optical properties of a sample with a higher resolution can be used to stimulate a growing interest in these instruments. Comparing with other common techniques such as scanning tunneling microscopy (STM) or atomic force microscopy (AFM), SNOM has less limitation in the kind of sample; it can be conducting or not conducting, transparent or opaque, soft or hard, thin or thick, and in air, in vacuum, or immersed in liquid. There is no direct mechanical interaction between probe and sample. The method is purely noncontact with all the advantages that this fact implies. The SNOM can be considered, simply speaking, the optical equivalent of the STM. Instead of electronic current, there is a flux of photons tunneling between the sample and a sharp transparent probe. The optical interaction between tip and sample is rather complex and not yet theoretically solved; however, it is generally accepted that the intensity of the collected optical field decreases dramatically with sample-probe separation, giving a high sensitivity to the system.

According to Abbe's theory of image formation [4], the resolving capability of an optical component is ultimately limited by the spreading out of each image point due to diffraction. Unless the aperture of the optical component is large enough to collect all the diffracted light, the finer aspects of the image will not correspond exactly to the object. The minimum resolution (*d*) for the optical component is thus limited by its aperture size and expressed by the Rayleigh criterion:

$$d = 0.61 \frac{\lambda_0}{NA} \qquad (1.12a)$$

Here, λ_0 is the wavelength in vacuum and *NA* is the numerical aperture for the optical component (usually 1.3–1.4 for modern objectives). Thus, the resolution limit is usually around $\lambda_0/2$ for conventional optical microscopy. Or in terms of the rigorous criterion for being able to resolve two objects in a light microscope, the resolution limit is

$$d > \frac{\lambda}{2\sin\theta} \qquad (1.12b)$$

Here, *d* is the distance between the two objects, λ is the wavelength of the incident light, and 2θ is the angle through which the light is collected, as show in Fig. 1.2.

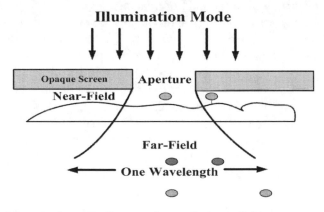

Figure 1.2 A schematic diagram of an optical near field.

According to Eq. 1.12b, the best resolution achievable with optical light is about 200 nm. With the introduction of NSOM, also known as SNOM, this limitation no longer exists, and optical resolution of <50 nm can be achieved.

Light passes through a subwavelength diameter aperture and illuminates a sample that is placed within its near field, at a distance much less than the wavelength of the light. The resolution achieved is far better than that which conventional optical microscopes can attain. In order to make an NSOM/SNOM experiment, a point light source must be brought near the surface that will be imaged (within nanometers). The point light source must then be scanned over the surface, without touching it, and the optical signal from the surface must be collected and detected.

This treatment only assumes the light diffracted into the far field that propagates without any restrictions. SNOM makes use of evanescent or nonpropagating fields that exist only near the surface of the object and occur when a light plane wave is incident on a medium interface in a condition of total internal reflection. An evanescent field, described by Eq. 1.13, is present on the opposite side and once coupled to a propagating medium (the probe tip) will produce a new real photon that can be detected.

$$E(x, z) = T_0 \exp\left(-i\omega t + ik_2 \frac{x}{n}\sin\theta_1\right)\exp\left(-k_2 z\sqrt{\frac{1}{n^2}\sin^2\theta_1 - 1}\right) \quad (1.13)$$

Here, n_1 and n_2 are the refractive indexes of mediums 1 and 2, respectively, $n = n_1/n_2$ is the relative refractive index, ω is the angular frequency, $k_2 = 2\pi/\lambda_2$ is the wave number, and λ_2 is the wavelength in medium 2.

These fields carry high-frequency spatial information about the object and have intensities that drop off exponentially with distance from the object. Because of this, the detector must be placed very close to the sample in the near-field zone, typically a few nanometers. As a result, near-field microscopy remains primarily a surface inspection technique. The detector is then rastered across the sample using a piezoelectric stage. The scanning can be done either at a constant height or with regulated height by using a feedback mechanism.

1.4   Quantum Confinement

Quantum confinement is the change of electronic and optical properties when the material sampled is of a sufficiently small size — typically 10 nm or less. The band gap increases as the size of the nanostructure decreases. Specifically, the phenomenon results from electrons and holes being squeezed into a dimension that approaches a critical quantum measurement, called the exciton Bohr radius. The first experimental evidence of the quantum confinement effects in clusters came from crystalline CuCl clusters grown in silicate glasses. Spectroscopic studies on these clusters clearly indicated an up to 0.1 eV blueshift of the absorption spectrum relative to the bulk. In the case of CdS clusters, the absorption threshold is observed to blueshift by up to 1 eV or more as the cluster size is decreased. When the size of the cluster is smaller, its band gap is larger. Consequently, the first absorption peak is shifted closer to the blue. A recent study of the X-ray absorption spectra in nanodiamond thin films with a grain diameter from 3.5 to 5 μm showed that the C 1s core exciton state and conduction band edge are shifted to higher energies with a decrease in the grain size, especially when the crystallite radius is smaller than ~1.8 nm. The conduction band of nanodiamonds with radius $R > 1.8$ nm when the crystalline contains more than 4,300 C atoms remains more or less bulk-like. Recently, Raty *et al.* [5] presented ab initio calculations based on density-functional theory (DFT) in order to investigate the quantum confinement effects in hydrogenated nanodiamonds. They detected a rapid decrease of the DFT energy gap from a value of 8.9 eV in methane to 4.3 eV in $C_8 7H_7 6$. The last value is very close to that of the bulk diamond (4.23 eV), obtained using the same method. This indicates that in contrast to Si and Ge, where quantum confinement effects persist up to 6–7 nm, in diamond there is no detectable quantum confinement for sizes larger than 1–1.2 nm. In addition, the authors predicted a slight influence of the surface structure reconstructed by hydrogen atoms on the optical properties.

An immediate optical feature of colloidal quantum dots is their coloration. While the material that makes up a quantum dot defines its intrinsic energy signature, the nanocrystal quantum-confined size is more significant at energies near the band gap. Thus, quantum dots of the same material but with different sizes can emit light of different colors. The physical reason is the quantum confinement effect. While the material has the larger dot, the atom emits lower energy in terms of fluorescence spectrum, called redder light emitting. Conversely, smaller dots emit bluer (higher-energy) light. The coloration is directly related to the energy levels of the quantum dot. Quantitatively speaking, the band gap energy that determines the energy (and hence color) of the fluorescent light is inversely proportional to the size of the quantum dot. Larger quantum dots have more energy levels that are also more closely spaced. This allows the quantum dot to absorb photons containing less energy, that is, those closer to the red end of the spectrum. Recent work has begun to suggest that the shape of the quantum dot may be a factor in the coloration as well, but not enough information is available yet [6]. Furthermore, it was shown that the lifetime of fluorescence is determined by the size of the quantum dot. Larger dots have more closely spaced energy levels in which the electron-hole pair can be trapped. Therefore, electron-hole pairs in larger dots live longer, causing larger dots to show a longer lifetime. As with any crystalline semiconductor, a quantum dot's electronic wave functions extend over the crystal lattice. Similar to a molecule, a quantum dot has both a quantized energy spectrum and a quantized density of electronic states near the edge of the band gap. Quantum dots can be synthesized with larger (thicker) shells (CdSe quantum dots with CdS shells). The shell thickness has shown direct correlation to the lifetime and emission intensity.

1.5 Photonic Crystal

The photonic crystal, or the photonic band gap, is the concept of the propagation confinement of photon in material that is the same as the electron energy band gap. The photonic band gap (PBG) is essentially the gap between the air line and the dielectric line in the dispersion relation of the PBG system. And photonic crystals are periodic optical nanostructures that are designed to affect the motion of photons in a similar way that periodicity of a semiconductor crystal affects the motion of electrons. Photonic crystals occur in nature and in various forms and have been studied scientifically for the last 100 years.

To design photonic crystal systems, it is essential to engineer the location and size of the band gap; this is done by computational modeling using any of the following methods:

- Plane wave expansion method
- Finite element method
- Finite-difference time-domain (FDTD) method
- Order-N spectral method
- KKR method

Essentially, these methods solve for the frequencies (normal models) of the photonic crystal for each value of the propagation direction given by the wave vector, or vice versa. The various lines in the band structure correspond to the different cases of **n**, the band index.

1.6   Finite Difference Time Domain Method

FDTD is a popular computational electrodynamics modeling technique. It is considered easy to understand and easy to implement in software. Since it is a time-domain method, solutions can cover a wide frequency range with a single simulation run. The FDTD method belongs in the general class of grid-based differential time-domain numerical modeling methods. The time-dependent Maxwell's equations (in partial differential form) are described using central-difference approximations to the space and time partial derivatives. The resulting finite-difference equations are solved in either software or hardware in a leapfrog manner: the electric field vector components in a volume of space are solved at a given instant in time; then the magnetic field vector components in the same spatial volume are solved at the next instant in time; and the process is repeated over and over again until the desired transient or steady-state electromagnetic field behavior is fully evolved.

FDTD techniques have emerged as the primary means to computationally model many scientific and engineering problems dealing with electromagnetic wave interactions with material structures. Current FDTD modeling applications range from near-DC (ultralow-frequency geophysics involving the entire Earth-ionosphere waveguide) through microwaves (radar signature technology, antennas, wireless communications devices, digital interconnects, and biomedical imaging/treatment) to visible light (photonic crystals, nanoplasmonics, solitons, and biophotonics).

When Maxwell's differential equations are examined, it can be seen that the change in the E field in time (the time derivative) is dependent on the change in the H field across space (the curl). This results in the basic FDTD time-stepping relation that, at any point in space, the updated value of the E field in time is dependent on the stored value of the E field and the numerical curl of the local distribution of the H field in space.

The H field is time-stepped in a similar manner. At any point in space, the updated value of the H field in time is dependent on the stored value of the H field and the numerical curl of the local distribution of the E field in space. Iterating the E field and H field updates results in a marching-in-time process wherein sampled-data analogs of the continuous electromagnetic waves under consideration propagate in a numerical grid stored in the computer memory.

Figure 1.3 is an illustration of a standard Cartesian Yee [7] cell used for FDTD, for which electric and magnetic field vector components are distributed. Visualized as a cubic voxel, the electric field components form the edges of the cube, and the magnetic field components form the normals to the faces of the cube. A three-dimensional (3-D) space lattice consists of a multiplicity of such Yee cells. An electromagnetic wave interaction structure is mapped into the space lattice by assigning appropriate values of permittivity to each electric field component and permeability to each magnetic field component.

This description holds true for 1-D, 2-D, and 3-D FDTD techniques. When multiple dimensions are considered, calculating the numerical curl can become complicated. Kane Yee's seminal 1966 paper proposed spatially staggering the vector components of the E field and H field about rectangular unit cells of a Cartesian computational grid so that each E-field vector component is located midway between a pair of H-field vector components, and conversely. This scheme, now known as a Yee lattice, has proven to be very robust and remains at the core of many current FDTD software constructs.

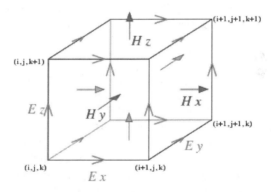

Figure 1.3   A standard Cartesian Yee cell used for FDTD.

Furthermore, Yee proposed a leapfrog scheme for marching in time wherein the E-field and H-field updates are staggered so that E-field updates are conducted midway during each time step between successive H-field updates, and conversely. On the plus side, this explicit time-stepping scheme prevents the need to solve simultaneous equations and, furthermore, yields dissipation-free numerical wave propagation. On the minus side, this scheme mandates an upper bound on the time step to ensure numerical stability. As a result, certain classes of simulations can require many thousands of time steps for completion.

In order to use FDTD, a computational domain must be established. The computational domain is simply the physical region over which the simulation will be performed. The E and H fields are determined at every point in space within that computational domain. The material of each cell within the computational domain must be specified. Typically, the material is either free-space (air), metal, or dielectric. Any material can be used as long as the permeability, permittivity, and conductivity are specified.

Once the computational domain and the grid materials are established, a source is specified. The source can be an impinging plane wave, a current on a wire, or an applied electric field, depending on the application. Since the E and H fields are determined directly, the output of the simulation is usually the E or H field at a point or a series of points within the computational domain. The simulation evolves the E and H fields forward in time. Processing may be done on the E and H fields returned by the simulation. Data processing may also occur while the simulation is ongoing. While the FDTD technique computes electromagnetic fields within a compact spatial region, scattered and/or radiated far fields can be obtained via near-to far-field transformations.

1.6.1   2-D FDTD Equations

The FDTD approach is based on a direct numerical solution of the time-dependent Maxwell's curl equations. The photonic device is laid out in the *X-Z* plane. The propagation is along *Z*. The *Y* direction is assumed to be infinite. This assumption removes all the $\partial/\partial y$ derivatives from Maxwell's equations and splits them into two (TE and TM) independent sets of equations. The 2-D computational domain is shown in Fig. 1.4. The space steps in the *X* and *Z* directions are Δx and Δz, respectively. Each mesh point is associated with a specific type of material and contains information about its properties such as refractive index and dispersion parameters.

1.6.1.1   TE waves

In the 2-D TE case (H_x, E_y, and H_z [nonzero components], propagation along *Z*, and the transverse field variations along *X*) in lossless media, Maxwell's equations take the following form:

$$\frac{\partial E_y}{\partial t} = \frac{1}{\varepsilon}\left(\frac{\partial H_x}{\partial z} - \frac{\partial H_z}{\partial x}\right), \quad \frac{\partial H_x}{\partial t} = \frac{1}{\mu_0}\frac{\partial E_y}{\partial z}, \quad \frac{\partial H_z}{\partial t} = \frac{1}{\mu_0}\frac{\partial E_y}{\partial x} \qquad (1.14)$$

Here, $\varepsilon = \varepsilon_0\varepsilon_r$ is the dielectric permittivity and μ_0 is the magnetic permeability of the vacuum. The refractive index is $n = \sqrt{\varepsilon_r}$. Each field is represented by a 2-D array — $E_y(i, k)$, $H_x(i, k)$, and $H_z(i, k)$ — corresponding to the 2-D mesh grid given in Fig. 1.3. The indices i and k account for the number of space steps in the X and Z direction, respectively. In the case of TE, the location of the fields in the mesh is shown in Fig. 1.4.

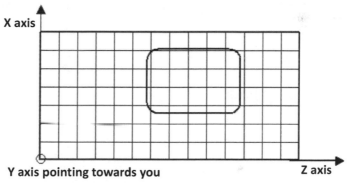

Figure 1.4 A numerical representation of the 2-D computational domain.

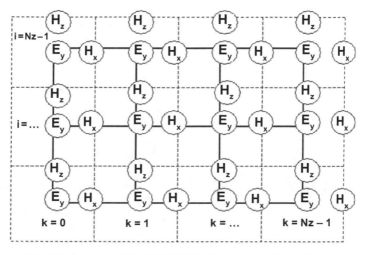

Figure 1.5 The location of the TE fields in the computational domain.

The TE fields stencil can be explained as follows. The E_y field locations coincide with the mesh nodes given in Fig. 1.4. In Fig. 1.5, the solid lines represent the mesh given in Fig. 1.4. The E_y field is considered to be the center of the FDTD space cell. The dashed lines form the FDTD cells. The magnetic fields H_x and H_z are associated with cell edges. The locations of the electric fields are associated with integer values of the indices i and k. The H_x field is associated with integer i and $(k + 0.5)$ indices. The H_z field is associated with $(i + 0.5)$ and integer k indices. The numerical analog in Eq. 1.14 can be derived from the following relation:

$$\frac{\partial E_y}{\partial t} = \frac{1}{\varepsilon}\left(\frac{\partial H_x}{\partial z} - \frac{\partial H_z}{\partial x}\right) \tag{1.15}$$

1.6.1.2 Numerical discretization

$$\frac{E_y^n(i,k) - E_y^{n-1}(i,k)}{\Delta t} = \frac{1}{\varepsilon}\frac{H_x^{n-1/2}(t,k+1/2) - H_x^{n-1/2}(i,k-1/2)}{\Delta z}$$

$$-\frac{1}{\varepsilon}\frac{H_z^{n-1/2}(i+1/2,k) - H_z^{n-1/2}(i-1/2,k)}{\Delta x} \tag{1.16}$$

The total set of numerical Eq. 1.16 takes the form

$$E_y^n(i,k) = E_y^{n-1}(i,k) + \frac{\Delta t}{\varepsilon\Delta z}\left[H_x^{n-1/2}(t,k+1/2) - H_x^{n-1/2}(i,k-1/2)\right]$$

$$-\frac{\Delta t}{\varepsilon\Delta z}\left[H_z^{n-1/2}(i+1/2,k) - H_z^{n-1/2}(i-1/2,k)\right]$$

$$H_x^{n+1/2}(i,k+1/2) = H_x^{n-1/2}(i,k+1/2) + \frac{\Delta t}{\mu_0\Delta z}\left[E_y^n(i,k+1) - E_y^n(i,k)\right] \tag{1.17}$$

$$H_z^{n+1/2}(i+1/2,k) = H_x^{n-1/2}(i+1/2,k) - \frac{\Delta t}{\mu_0\Delta x}\left[E_y^n(i+1,k) - E_y^n(i,k)\right] \tag{1.18}$$

The superscript n labels the time steps, while the indices i and k label the space steps and Δx and Δz along the x and z directions, respectively. This is the so-called Yee numerical scheme applied to the 2-D TE case. It uses central difference approximations for the numerical derivatives in space and time, both having second-order accuracy. The sampling in space is on a subwavelength scale. Typically, 10 to 20 steps per wavelength are needed. The sampling in time is selected to ensure numerical stability of the algorithm. The time step is determined by the Courant limit:

$$\Delta t \le 1 \Big/ \Big(c \sqrt{(\Delta x)^2 + 1/(\Delta z)^2} \Big) \tag{1.19}$$

1.6.1.3 TM waves

In the 2-D TM case (E_x, H_y, and E_z [nonzero components], propagation along Z, and transverse field variations along X) in lossless media, Maxwell's equations take the following form:

$$\frac{\partial H_y}{\partial t} = -\frac{1}{\mu_0}\left(\frac{\partial E_x}{\partial z} - \frac{\partial E_z}{\partial x}\right), \ \frac{\partial E_x}{\partial t} = \frac{1}{\varepsilon}\frac{\partial H_y}{\partial z}, \ \frac{\partial E_z}{\partial t} = \frac{1}{\varepsilon}\frac{\partial H_y}{\partial x} \tag{1.20}$$

The location of the TM fields in the computational domain follows the same philosophy and is shown in Fig. 1.6.

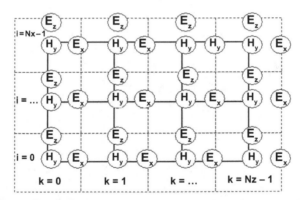

Figure 1.6 The location of the TM fields in the computational domain.

Now, the electric field components E_x and E_z are associated with the cell edges, while the magnetic field H_y is located at the cell center. The TM algorithm can be presented in a way similar to Eq. 1.18.

1.6.2 3-D FDTD Equations

In 3-D simulations, the simulation domain is a cubic box and the space steps are D_x, D_y, and D_z in x, y, and z directions, respectively. Each field components is presented by a 3-D array — $E_x(i,j,k)$, $E_y(i,j,k)$, $E_z(i,j,k)$, $H_x(i,j,k)$, $H_y(i,j,k)$, and $H_z(i,j,k)$. The field components' positions in Yee's cell are shown in Fig. 1.7. These placements and the notation show that the E and H components are interleaved at intervals of in space and for the purpose of implementing a leapfrog algorithm.

1.6.2.1 Output data

The fields propagated by the FDTD algorithm are the time-domain fields. At each location of the computational domain, they have a form similar to that given in Eq. 1.21.

$$E_y(x, z) = BG(x, z)\sin(\omega t + \varphi_i) \qquad (1.21)$$

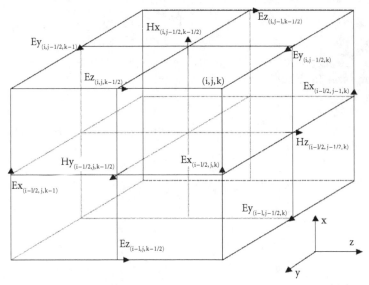

Figure 1.7 Displacement of the electric and magnetic field vector components about a cubic unit cell of the Yee space lattice.

Here, B is the amplitude of the field at that particular location, G is the wave profile, and φ_i is the corresponding phase. However, the values of B and φ_i are not accessible from the time-domain field values. To get the full amplitude/phase wave information, we need the stationary complex fields that correspond to the waveform Eq. 1.20. The complex fields are the source of all useful information, such as output and reflected powers and overlap integrals with modal fields. Those complex fields are calculated by a run-time Fourier transform performed in the last time period of the simulation. The final complex fields can be visualized at specific output planes located properly in the computational domain. The output plane located on the right side of the input plane is in the total field region and collects the total field data. The output plane on the left of the input plane is in the reflected field region and collects reflected field data. It is also called the reflection plane. Output planes can also be situated along the Z axis.

1.7 Light Pulse in the Ring Resonator

The basic configuration of an add/drop ring resonator with radius *r* and a waveguide is described in Fig. 1.8. A single unidirectional mode of the resonator is excited, the coupling is lossless, single polarization is considered, none of the waveguide segments and coupler elements couple waves of different polarization, and the various kinds of losses occurring along the propagation of light in the ring resonator filter are incorporated in the attenuation constant. The interaction can be described by the following equations:

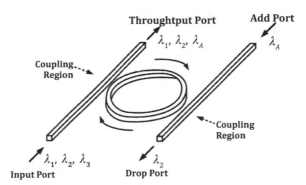

Figure 1.8   A ring resonator add/drop filter.

$$E_{th} = \sqrt{1-\gamma_1}\left(\sqrt{1-\kappa_1}E_{in} + j\sqrt{\kappa_1}E_4\right)$$ (1.22)

$$E_1 = \sqrt{1-\gamma_1}\left(\sqrt{1-\kappa_1}E_4 + j\sqrt{\kappa_1}E_{in}\right)$$ (1.23)

$$E_3 = \sqrt{1-\gamma_2}\left(\sqrt{1-\kappa_2}E_2 + j\sqrt{\kappa_2}E_{add}\right)$$ (1.24)

$$E_{drop} = \sqrt{1-\gamma_2}\left(\sqrt{1-\kappa_2}\,E_{add} + j\sqrt{\kappa_2}\,E_2\right) \qquad (1.25)$$

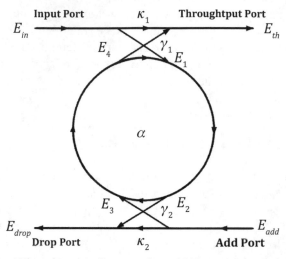

Figure 1.9   The schematic diagram of an add/drop ring resonator.

Here, E_{in}, E_{th}, E_{add}, and E_{drop} are the electric fields at the input port, throughput port, add port, and drop port respectively, κ_1 and κ_2 are the coupling coefficients, γ_1 and γ_2 are the fractional coupler intensity losses, and j refers to the complex number where $j^2 = -1$. The electric fields inside the ring resonator E_1, E_2, E_3, and E_4 are described by

$$E_2 = E_1 e^{\frac{-\alpha}{2}L - jk_n L} \qquad (1.26)$$

$$E_4 = E_3 e^{\frac{-\alpha}{2}L - jk_n L} \qquad (1.27)$$

Here α is the attenuation per unit length (zero attenuation: $\alpha = 0$); k_n is the effective wave number, and it's related to the wavelength λ through $k_n = 2\pi n_{eff}/\lambda$, where n_{eff} is the effective refractive index of the medium; and $L = 2\pi R$ is the circumference of the ring with the radius R.

The complex mode amplitudes E are normalized, so that their squared magnitude corresponds to the modal power. According to Fig. 1.9 and Eqs. 1.22–1.27, assuming that there is no electric field at the add port, the intensities at the throughput port and the drop port can be described by

$$\left|\frac{E_{th}}{E_{in}}\right|^2 = \frac{(1-\kappa_1)-2\sqrt{1-\kappa_1}\cdot\sqrt{1-\kappa_2}\cdot e^{\frac{-\alpha}{2}L}\cos(k_n L)+(1-\kappa_2)e^{-\alpha L}}{1+(1-\kappa_1)(1-\kappa_2)e^{-\alpha L}-2\sqrt{1-\kappa_1}\cdot\sqrt{1-\kappa_2}e^{\frac{-\alpha}{2}L}\cos(k_n L)}$$

(1.28)

$$\left|\frac{E_{drop}}{E_{in}}\right|^2 = \frac{\kappa_1\kappa_2\cdot e^{\frac{-\alpha}{2}L}}{1+(1-\kappa_1)(1-\kappa_2)e^{-\alpha L}-2\sqrt{1-\kappa_1}\cdot\sqrt{1-\kappa_2}e^{\frac{-\alpha}{2}L}\cos(k_n L)}$$

(1.29)

1.8 Conclusion

The basic background of the nanophotonics is presented, where both theoretical background and simulation method using OptiWave FDTD are described, which can be used to support the following chapters.

References

1. E. H. Synge, "A suggested method for extending the microscopic resolution into the ultramicroscopic region," *Phil. Mag.*, **6**, 356 (1928).

2. D. Pohl and D. Courjon (eds), "Near-field optics," *NATO ASI Ser.*, **E242**, 121–123, Kluwer, Dordrecht (1993).

3. C. Bainier, D. Courjon, F. Baida, and C. Girard, "Evanescent inter-ferometry by scanning optical tunneling detection," *J. Opt. Soc. Am.*, **13**, 267–275 (1996).

4. E. Abbe, "Beiträge zur theorie des mikroskops und der mikroskopischen wahrnehmung," *Archiv für Anatomie*, **9**(1), 413–468 (1873).

5. J. Y. Raty, G. Galli, C. Bostedt, T. W. van Buuren, and L. J. Terminello, "Quantum confinement and Fullerenelike surface reconstructions in nanodiamonds," *Phys. Rev. Lett.*, **90**, 037401 (2003).

6. A. F. Van Driel, "Frequency-dependent spontaneous emission rate from CdSe and CdTe nanocrystals: influence of dark states," *Phys. Rev. Lett.*, **95**(23), 236804 (2005).

7. K. Yee, "Numerical solution of initial boundary value problems involving Maxwell's equations in isotropic media," *IEEE Trans. Antennas Propag.*, **14**, 302–307 (1966).

Chapter 2

Optical Cryptography

2.1   Introduction

A PANDA ring resonator has been successfully used to investigate the dynamic behavior of dark-bright soliton collision within a modified add/drop filter [1–4]. Optical devices have become an integrated component in advanced optical technology applications and are being widely used in optical communication, optical sensing, signal processing, etc. Therefore, the communication security segment has been recognized as a promising tool for information security and privacy, which is largely required in world networks.

Today, security schemes, such as quantum and optical techniques, are being widely used in many applications, such as sensing [5], computing [6], communication [7], and signal processing [8]. Optical devices are especially being used for security communications [9–11]. Recently, the use of a device known as a ring resonator, specifically, a microring resonator in the form of an optical add/drop filter, has been found in many applications [12–14]. In these applications, transmitted signals could be suppressed with chaotic signals and the required signals could be retrieved by the add/drop filter and the encryption-decryption method by using the proposed optical design system. However, the search for new devices and techniques is an ongoing process.

In this chapter, we propose the use of key suppression and recovery for optical cryptography by using a PANDA ring resonator for high-security communication. The required key can be suppressed

Nanophotonics: Devices, Circuits, and Systems
Preecha P. Yupapin, Keerayoot Srinuanjan, and Surachart Kamoldilok
Copyright © 2013 Pan Stanford Publishing Pte. Ltd.
ISBN 978-981-4364-36-2 (Hardcover), 978-981-4364-37-9 (eBook)
www.panstanford.com

(buried) by the noisy signals and the required signals can be secured and recovered (retrieved) by the specifically designed optical device. The required data can be encrypted and decrypted by the optical encryption-decryption keys, respectively, and both keys can be generated by using the suppressed optical keys (LIP signals). This method can be used to form secure communication in either digital or analog communications.

2.2 Model of an Optical Cryptography System

The proposed system consists of three optical devices, which are a PANDA ring resonator, an add/drop filter, and a Mach-Zehnder interferometer as shown in Figs. 2.1–2.3. The transmitter part of an optical cryptography system consists of a PANDA ring resonator, two add/drop filters, and a Mach-Zehnder interferometer. The receiver part consists of two add/drop filters and a Mach-Zehnder interferometer.

An add/drop filter and the double microring resonators, together known as a PANDA ring resonator, are shown in Fig. 2.1. To perform the dark-bright soliton conversion, the dark and bright solitons are input into the add/drop optical filter. The input optical field (E_{in}) and the control port optical field (E_{con}) of the bright-dark soliton pulses are given by Eqs. 2.1 and 2.2 [15].

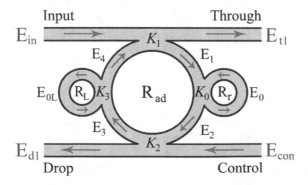

Figure 2.1   A schematic diagram of a PANDA ring resonator.

$$E_{in}(t) = A\operatorname{sech}\left[\frac{T}{T_0}\right]\exp\left[\left(\frac{x}{2L_D}\right) - i\phi(t)\right] \qquad (2.1)$$

and

$$E_{con}(t) = A\tanh\left[\frac{T}{T_0}\right]\exp\left[\left(\frac{x}{2L_D}\right) - i\phi(t)\right] \qquad (2.2)$$

Here A and z are the optical field amplitude and propagation distance, respectively. $T = t - \beta_1 z$, where T is the propagation time of a soliton pulse in a frame moving at the group velocity, and $L_D = T_0^2/|\beta_2|$ is the dispersion length of the soliton pulse. β_1 and β_2 are the coefficients of the linear and second-order terms of the Taylor expansion of the propagation constant. In Eqs. 2.1 and 2.2, T_0 is the soliton pulse propagation time (or the soliton pulse width) at initial input, t the soliton phase shift time, and ω_0 the angular frequency shift of the soliton. This solution describes a pulse that keeps its temporal width invariance as it propagates and is, therefore, called a temporal soliton. When a soliton peak intensity $\left(|\beta_2/\Gamma T_0^2|\right)$ is given, then T_0 is known. For a soliton pulse in a microring device, a balance should be achieved between the dispersion length (L_D) and the nonlinear length ($L_{NL} = 1/\Gamma\phi_{NL}$), where $\Gamma = n_2 k_n$ is the length scale over which dispersive or nonlinear effects make the beam wider or narrower for the soliton pulse $L_D = L_{NL}$.

The PANDA ring resonator shown in Fig. 2.1 is used for the binary coded suppression. When the input light pulse passes through the first optical coupler of the add/drop optical multiplexing system, the transmitted and circulated optical fields can be written as follows [16]:

$$E_{t1} = \sqrt{1-\gamma_1}\left[\sqrt{1-\kappa_1}\,E_{in} + j\sqrt{\kappa_1}E_4\right] \qquad (2.3)$$

$$E_1 = \sqrt{1-\gamma_1}\left[\sqrt{1-\kappa_1}\,E_4 + j\sqrt{\kappa_1}E_{in}\right] \qquad (2.4)$$

$$E_2 = E_0 E_1 e^{-\frac{\alpha L}{2 2} - jk_n\frac{L}{2}} \qquad (2.5)$$

Here κ_1 is the intensity coupling coefficient, γ_1 the fractional coupler intensity loss, α the attenuation coefficient, $k_n = 2\pi/\lambda$ the wave propagation number, λ the input wavelength light field, and $L = 2\pi/R_{ad}$, where R_{ad} is the radius of the add/drop filter.

The optical fields at the second coupler of the add/drop optical multiplexing system are given by

$$E_{d1} = \sqrt{1-\gamma_2}\left[\sqrt{1-\kappa_2}\,E_{con} + j\sqrt{\kappa_2}E_2\right] \qquad (2.6)$$

$$E_3 = \sqrt{1-\gamma_2}\left[\sqrt{1-\kappa_2}\,E_2 + j\sqrt{\kappa_2}\,E_{con}\right]$$ (2.7)

$$E_4 = E_{0L}E_3 e^{-\frac{\alpha L}{2}\frac{L}{2}-jk_n\frac{L}{2}}$$ (2.8)

Here κ_2 is the intensity coupling coefficient and γ_2 is the fractional coupler intensity loss. E_0 and E_{0L} are the circulated light field components of the microrings with radii R_r and R_L, which couple into the right hand side (RHS) and left hand side (LHS) of the add/drop optical multiplexing system, respectively. The transmitted and circulated components of the light field in the R_r are given by

$$E_2 = \sqrt{1-\gamma}\left[\sqrt{1-\kappa_0}\,E_1 + j\sqrt{\kappa_0}\,E_{r2}\right]$$ (2.9)

$$E_{r1} = \sqrt{1-\gamma}\left[\sqrt{1-\kappa_0}\,E_{r2} + j\sqrt{\kappa_0}\,E_1\right]$$ (2.10)

$$E_{r2} = E_{r1}e^{-\frac{\alpha}{2}L_1-jk_nL_1}$$ (2.11)

Here κ_0 is the intensity coupling coefficient, γ the fractional coupler intensity loss, α the attenuation coefficient, $k_n = 2\pi/\lambda$ the wave propagation number, λ the input wavelength of the light field, and $L_1 = 2\pi R_r$, where R_r is the radius of right microring.

The circulated roundtrip light fields, E_{r1} and E_{r2}, of the RHS microring given in Eqs. 2.12 and 2.13, respectively, are derived from Eqs. 2.9–2.11.

$$E_{r1} = \frac{j\sqrt{1-\gamma}\sqrt{\kappa_0}\,E_1}{1-\sqrt{1-\gamma}\sqrt{1-\kappa_0}\,e^{-\frac{\alpha}{2}L_1-jk_nL_1}}$$ (2.12)

$$E_{r2} = \frac{j\sqrt{1-\gamma}\sqrt{\kappa_0}\,E_1 e^{-\frac{\alpha}{2}L_1-jk_nL_1}}{1-\sqrt{1-\gamma}\sqrt{1-\kappa_0}\,e^{-\frac{\alpha}{2}L_1-jk_nL_1}}$$ (2.13)

Thus, the output circulated light field, E_0, for the right microring is given by

$$E_0 = E_1\left\{\frac{\sqrt{(1-\gamma)(1-\kappa_0)}-(1-\gamma)e^{-\frac{\alpha}{2}L_1-jk_nL_1}}{1-\sqrt{(1-\gamma)(1-\kappa_0)}\,e^{-\frac{\alpha}{2}L_1-jk_nL_1}}\right\}$$ (2.14)

Similarly, the output circulated light field, E_{0L}, for the left micro-ring of the add/drop optical multiplexing system is given by

$$E_{0L} = E_3 \left\{ \frac{\sqrt{(1-\gamma_3)(1-\kappa_3)} - (1-\gamma_3)e^{-\frac{\alpha}{2}L_2 - jk_nL_2}}{1 - \sqrt{(1-\gamma_3)(1-\kappa_3)}e^{-\frac{\alpha}{2}L_2 - jk_nL_2}} \right\} \tag{2.15}$$

where κ_3 is the intensity coupling coefficient, γ_3 the fractional coupler intensity loss, α the attenuation coefficient, $k_n = 2\pi/\lambda$ the wave propagation number, λ the input wavelength light field, and $L_2 = 2\pi R_L$, where R_L is the radius of LHS microring.

From Eqs. 2.3–2.15, the circulated light fields E_1, E_3, and E_4 can be derived, given that $x_1 = (1 - \gamma_1)^{1/2}$, $x_2 = (1 - \gamma_2)^{1/2}$, $y_1 = (1 - \kappa_1)^{1/2}$, and $y_2 = (1 - \kappa_2)^{1/2}$. Thus,

$$E_1 = \frac{jx_1\sqrt{\kappa_1}E_{in} + jx_1x_2y_1\sqrt{\kappa_2}E_{0L}E_{con}e^{-\frac{\alpha L}{2\,2} - jk_n\frac{L}{2}}}{1 - x_1x_2y_1y_2E_0E_{0L}e^{-\frac{\alpha}{2}L - jk_nL}} \tag{2.16}$$

$$E_3 = x_2y_2E_0E_1e^{-\frac{\alpha l}{2\,2} - jk_n\frac{l}{2}} + jx_2\sqrt{\kappa_2}E_{con} \tag{2.17}$$

$$E_4 = x_2y_2E_0E_{0L}E_1e^{-\frac{\alpha}{2}L - jk_nL} + jx_2\sqrt{\kappa_2}E_{0L}E_{con}e^{-\frac{\alpha L}{2\,2} - jk_n\frac{L}{2}} \tag{2.18}$$

From Eqs. 2.3, 2.5, and 2.16–2.18, the output optical field of the through port (E_{t1}) is expressed by

$$E_{t1} = x_1y_1k_{in} + \left(\begin{array}{c} jx_1x_2y_2\sqrt{\kappa_1}E_0E_{0L}E_1 \\ -x_1x_2\sqrt{\kappa_1\kappa_2}E_{0L}E_{t2} \end{array} \right)e^{-\frac{\alpha L}{2\,2} - jk_n\frac{L}{2}} \tag{2.19}$$

The power output of the through port (P_{t1}) is written by

$$P_{t1} = (E_{t1}) \cdot (E_{t1})^* = |E_{t1}|^2 \tag{2.20}$$

Similarly, from Eqs. 2.5, 2.6, 2.16–2.18, the output optical field of the drop port (E_{d1}) is given by

$$E_{d1} = x_2y_2E_{con} + jx_2\sqrt{\kappa_2}E_0E_1e^{-\frac{\alpha L}{2\,2} - jk_n\frac{L}{2}} \tag{2.21}$$

The power output of the drop port (P_{d1}) is expressed by

$$P_{d1} = (E_{d1}) \cdot (E_{d1})^* = |E_{d1}|^2 \tag{2.22}$$

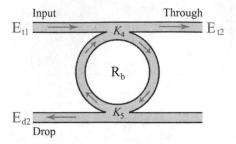

Figure 2.2   A schematic diagram of an add/drop filter.

An add/drop optical filter with the appropriate parameters are shown in Fig. 2.2. The electric field detected by photodetector is given by the following equation [17]:

$$E_{t2} = E_{t1} \frac{-\sqrt{1-\kappa_4}\,e^{-\frac{\alpha}{2}L_b - jk_n L_b} + \sqrt{1-\kappa_4}}{1 - \sqrt{1-\kappa_4}\sqrt{1-\kappa_5}\,e^{-\frac{\alpha}{2}L_b - jk_n L_b}} \tag{2.23}$$

Here $L_b = 2\pi R_b$, where R_b is the radius of the add/drop optical filter decoded as shown in Fig. 2.2. The power output of the drop port (P_{t2}) is expressed by

$$P_{t2} = (E_{t2}) \cdot (F_{t2})^* = |E_{t2}|^2 \tag{2.24}$$

The electric field detected by photodetector is given by

$$E_{d2} = E_{t1} \frac{-\sqrt{\kappa_4 \kappa_5}\,e^{-\frac{\alpha}{2}\frac{L_b}{2} - jk_n \frac{L_b}{2}}}{1 - \sqrt{1-\kappa_4}\sqrt{1-\kappa_5}\,e^{-\frac{\alpha}{2}L_b - jk_n L_b}} \tag{2.25}$$

The power output of the drop port (P_{d2}) is expressed by

$$P_{d2} = (E_{d2}) \cdot (E_{d2})^* = |E_{d2}|^2 \tag{2.26}$$

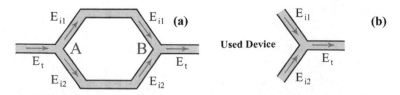

Figure 2.3   A schematic diagram of Mach-Zehnder interferometer.

The proposed system uses the Mach-Zehnder interferometer shown in Fig. 2.3, in which the required optical cryptography is performed incorporating the Mach-Zehnder interferometer. Considering the output (E_t) at point B, which is equal to input 1 (E_{i1}) plus input 2 (E_{i2}), the electric field detected by a photodetector is given by the following equation [18]:

$$E_t = E_{i1} + E_{i2} \tag{2.27}$$

The power output of the drop port (P_{d2}) is expressed by

$$P_t = (E_t) \cdot (E_t)^* = |E_t|^2 \tag{2.28}$$

2.3   Key Suppression and Recovery

In the simulation for optical key suppression, the used parameters of a PANDA ring resonator are fixed as $\kappa_0 = 0.1, \kappa_1 = 0.2, \kappa_2 = 0.2$, and $\kappa_3 = 0.1$. The ring radii are $R_{ad} = 200$ μm, $R_r = 15$ μm, and $R_L = 15$ μm. A_{eff} for the PANDA ring resonator and the right and left microring resonators are 0.50 μm², 0.25 μm², and 0.25 μm² [19], respectively. For optical key recovery, the parameters of the add/drop filter are fixed as $\kappa_4 = 0.5, \kappa_5 = 0.2, R = 100$ μm, and $A_{eff} = 0.25$ μm². Ideally, an optical key suppression and recovery system should be fabricable, which can conform to the parameters of the practical devices. Simulation results of the optical key signal with center wavelengths are at $\lambda_0 = 1.50$ μm.

Figures 2.4a–d show the simulation results of the optical key suppression for $|E_{in}|^2$, $|E_{con}|^2$, $|E_{t1}|^2$, and $|E_{d1}|^2$, with $R_r = 15$ μm, $R_L = 15$ μm, $R_{ad} = 200$ μm, and $\alpha = 5 \times 10^{-5}$ dBmm^{-1}. Figure 2.4a is the input port for optical key suppression. A bright soliton pulse with 1 W peak power is input into this port. Figure 2.4b is the control port output that uses a dark soliton pulse with 1 W peak power. The power output of the drop port is shown in Fig. 2.4d. Figure 2.4c shows the power outputs of the through port, which are the optical key suppression signals that can be transmitted to the receiver for secure optical communication and can be used as reference signals in communication. Moreover, the peak power outputs of the through port and drop port are 2.3 W and 3.2 W, respectively. They are larger than the input light pulse due to the optical nonlinear effects.

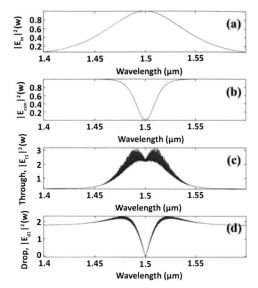

Figure 2.4   Simulation results for optical key suppression, where (a) $|E_{in}|^2$, (b) $|E_{con}|^2$, (c) $|E_{t1}|^2$, and (d) $|E_{d1}|^2$ with $R_r = 15$ μm, $R_L = 15$ μm, $R_{ad} = 200$ μm, and $\alpha = 5 \times 10^{-5}$ dBmm^{-1}.

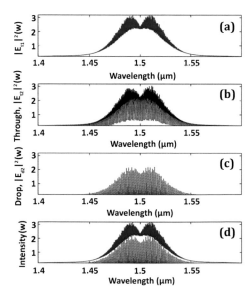

Figure 2.5   Simulation results for optical key recovery, where (a) $|E_{t1}|^2$, (b) $|E_{t2}|^2$, (c) $|E_{d2}|^2$ and (d) comparison of key suppression and key recovery, where $R_b = 100$ μm and $\alpha = 5 \times 10^{-5}$ dBmm^{-1}.

Figures 2.5a–c show the simulation result of the optical key recovery for $|E_{t1}|^2$, $|E_{t2}|^2$, and $|E_{d2}|^2$, respectively and Fig. 2.5d shows a comparison of the suppression and recovery keys, with R_b = 30 μm and α = 5 × 10^{-5} dBmm^{-1}. Here Fig. 2.5a shows the input port for the optical recovery key. The suppression key signal, which looks like a noisy signal, is input into this port that produces a highly secure optical communication. Figures 2.5b, c show the power output of the through and drop ports for signal recovery, respectively. The power output from the drop port is the analog signal by a sender and is used as an optical key in a cryptography system, which is also known as the LIP signal (key). Figure 2.5d shows a comparison of the suppressed optical key (*blue line*) and the recovered optical key (*red line*) in which secret signals are hidden by noisy signals.

2.4 Optical Cryptography System

Figure 2.6 shows the schematic diagram of an optical cryptography system, where PA is the PANDA ring resonator, AD the add/drop filter, and MZ the Mach-Zehnder interferometer. The transmitter part consists of one PA, two ADs, and one MZ. A bright soliton pulse (E_{in}) and a dark soliton pulse (E_{con}) is input into the input port and the control port, respectively, of the system (key suppression part) for key suppression. The output signal obtained is E_{t1}. It is the optical suppressed key that is sent to the receiver as the security signal as shown in Fig. 2.4c. First, the AD generates the optical key, or the LIP key, at the transmitter side (E_{d2}) from the suppressed signal as shown in Fig. 2.5c. Second, an AD function is generated by the LIP key to form the encrypted and decrypted keys. However, the transmitter uses the encryption key from the encryption data by MZ. This means that the data that is encrypted is sent to the required receiver, where finally the LIP key from the recovery key becomes the ciphertext.

The receiver part consists of two ADs and one MZ. The signal (E_{t1}) is sent to the input port by the transmitter as shown in Fig. 2.5a. The output signal (E_{d2}) that departs from the drop port is the LIP key as shown in Fig. 2.5c, which is sent by the transmitter. The LIP key is used for encrypted and decrypted key generations. However, the receiver part uses the decryption key from the data encryption by MZ. The decryption key is used for ciphertext decryption, which is also sent by the transmitter. Thus, our proposed system can be claimed as a new security technique that uses the optical cryptography design for the transmission of secret data that can be analog or digital data signals. Moreover, triple security functions can be realized by using the suppressed optical key, by changing the optical key in every data frame and by using the new optical cryptography technique. Figures 2.7a–d show the simulation result of optical cryptography at the sender side. Figure 2.7a is an example of data signal (E_d). Figure 2.7b shows the LIP key, which is generated by the key recovery part. Figure 2.7c shows the encryption key, which is generated by the LIP key using the add/drop filter. Figure 2.7d shows the optical ciphertext from which the encrypted key originates to encrypt the data signal.

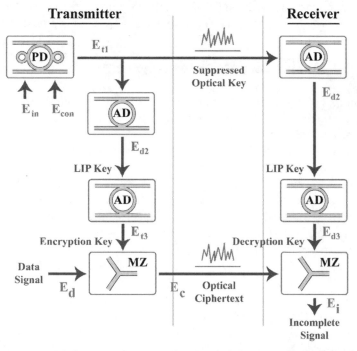

Figure 2.6   A schematic diagram of an optical cryptography system (PA: PANDA ring resonator, AD: Add/drop filter, and MZ: Mach-Zehnder interferometer).

Figures 2.8a–d show the simulation result of optical cryptography at the receiver side. Figure 2.8a shows the optical ciphertext, which is sent by the transmitter. Figure 2.8b shows LIP key, which is generated by the key recovery part. Figure 2.8c shows the decryption key, which is generated by the LIP key (E_{t2}) by using the add/drop filter. Figure 2.8d shows the incomplete data (E_i) signal, which generates the

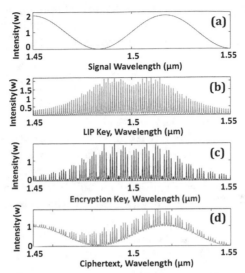

Figure 2.7   Simulation results of optical cryptography at sender side, where (a) data signal, (b) LIP signal (key), (c) encryption key, and (d) ciphertext.

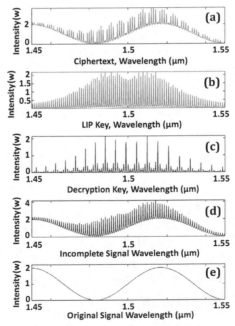

Figure 2.8   Simulation results of optical cryptography at receiver side, where (a) ciphertext, (b) LIP signal (key), (c) decryption key, (d) incomplete data signal, and (e) original data signal.

decryption key to decrypt the ciphertext signal. Figure 2.8e shows how the receiver can finally get the original signal (secret signal) by using the original data (E_d) = incomplete data (E_i) – LIP key (E_{t2}).

2.5 Conclusion

In this chapter a design for optical communication security was proposed by using the key suppression and recovery methods and was named as the optical cryptography system. By using the dark-bright soliton pair within a PANDA ring resonator. By using the practical parameters, the proposed system can be checked for compliance to possible fabrications. The key suppression is designed by using a PANDA ring resonator and the key recovery is obtained by using the add/drop filter. The optical cryptography system is designed by using the add/drop filter and Mach-Zehnder interferometer concept. Therefore, the add/drop filter can be used to generate the encryption and decryption keys from the LIP key. The Mach-Zehnder interferometer concept is used for signal combination (i.e., interference). The security functions of the proposed optical cryptography system can be realized by achieving security by changing (rearranging) the suppressed key in every data frame. The simulation results obtained show that the proposed system can indeed be used to achieve key suppression and recovery for an optical cryptography system.

References

1. N. Suwanpayak, M. A. Jalil, C. Teeka, J. Ali, and P. P. Yupapin, "Optical vortices generated by a PANDA ring resonator for drug trapping and delivery applications," *Biomed. Opt. Express*, **2**(1), 159–168 (2011).

2. P. P. Yupapin, N. Suwanpayak, B. Jukgoljun, and C. Teeka, "Hybrid transceiver using a PANDA ring resonator for nanocommunication," *Phys. Express*, **1**(1), 1–9 (2011).

3. M. Tasakorn, C. Teeka, R. Jomtarak, and P. P. Yupapin, "Multitweezers generation control within a nanoring resonator system," *Opt. Eng.*, **49**(7), 75002 (2010).

4. B. Jukgoljun, N. Suwanpayak, C. Teeka, and P. P. Yupapin, "Hybrid transceiver and repeater using a PANDA ring resonator for nanocommunication," *Opt. Eng.*, **49**(12), 125003 (2010).

5. P. Hua, B. J. Luff, G. R. Quigley, J. S. Wilkinson, and K. Kawaguchi, "Integrated optical dual Mach-Zehnder interferometer sensor," *Sens. Actuators B*, **87**, 250–257 (2002).

6. C. Kostrzewa, R. Moosburger, G. Fischbeck, B. Schuppert, and K. Petermann, "Tunable polymer optical add/drop filter for multi-wavelength networks," *IEEE Photon. Technol. Lett.*, **9**(11), 1487–1489 (1997).

7. P. D. Townsend, "Quantum cryptography on optical fiber networks," *Opt. Fiber Technol.*, **4**(4), 345–370 (1998).

8. T. Carmon, T. J. Kippenberg, L. Yang, H. Rokhsari, S. Spillane, and K. J. Vahala, "Feedback control of ultra-high-Q microcavities: Application to micro-Raman lasers and microparametric oscillators," *Opt. Express*, **13**(9), 3558–3566 (2005).

9. W. Siririth, S. Mitatha, O. Pingern, and P. P. Yupapin, "A novel temporal dark-bright solitons conversion system via an add/drop filter for signal security use," *Optik*, **121**(21), 1955–1958 (2010).

10. B. Knobnob, S. Mitatha, K. Dejhan, S. Chaiyasoonthorn, and P. P. Yupapin, "Dark–bright optical solitons conversion via an optical add/drop filter for signals and networks security applications," *Optik*, **121**(19), 1743–1747 (2010).

11. P. P. Yupapin, "Generalized quantum key distribution via micro ring resonator for mobile telephone networks," *Optik*, **121**(5), 422–425 (2010).

12. P. Gallion, F. Mendieta, and S. Jiang, "Signal and quantum noise in optical communications and cryptography," *Prog. Opt.*, **52**, 149–259 (2009).

13. Y. Dumeige, C. Arnaud, and P. Féron, "Combining FDTD with coupled mode theories for bistability in micro-ring resonator," *Opt. Commun.*, **250**(4–6), 376–383 (2005).

14. P. Rabiei, "Calculation of losses in micro-ring resonators with arbitrary refractive index or shape profile and its applications," *J. Lightwave Technol.*, **23**(3), 1295–1301 (2005).

15. K. Sarapat, N. Sangwara, K. Srinuanjan, P. P. Yupapin and N. Pornsuwancharoen, "Novel dark-bright optical solitons conversion system and power amplification," *Opt. Eng.*, **48**(4), 045004-1-7 (2009).

16. T. Phatharaworamet, C. Teeka, R. Jomtarak, S. Mitatha, and P. P. Yupapin, "Random binary code generation using dark-bright soliton conversion control within a PANDA ring resonator," *J. Lightwave Technol.*, **28**(19), 2804–2809 (2010).

17. D. G. Rabus, M. Hamacher, U. Troppenz, and H. Heidrich, "Optical filters based on ring resonators with integrated semiconductor optical amplifiers In GaInAsP–InP," *IEEE J. Sel. Top. Quantum Electron.*, **8**(6), 1405–1411 (2002).

18. A. Srivastava and S. Medhekar, "Switching of one beam by another in a Kerr type nonlinear Mach-Zehnder interferometer," *Opt. Laser Technol.*, **43**(1), 29–35 (2006).

19. Y. Kokubun, Y. Hatakeyama, M. Ogata, S. Suzuki and N. Zaizen, "Fabrication technologies for vertically coupled microring resonator with multilevel crossing busline and ultracompact-ring radius," *IEEE J. Sel. Top. Quantum Electron.*, **11**, 4–10 (2005).

Chapter 3

All-Optical Adder/Subtractor

3.1   Introduction

Optical devices have become important components in the new era of merging research areas in which the use of small-scale optical device is involved in the research area known as nanotechnology, where more interesting applications have appeared, especially, when the device dimension has reached the nanoscale regime. Many research works have shown the potential of optical devices in future applications, one of them being the logical device. This device is recognized as an important part of a computer in which not only the electronic circuit is operating improperly due to ultra-high speed signal processing requirement but bandwidth is also a limitation. Nowadays, the use of all-optical signal processing has increased significantly. Although various architectures, algorithms, and logical and arithmetic operations have been proposed, such as systems of semiconductor optical amplifier (SOA) [1–4], a quantum dot [5, 6], a terahertz optical asymmetric demultiplex (TOAD) [7, 8], cascaded microring resonators [9], an all-optical arithmetic unit [10, 11], an all-optical binary counter [12], and an all-optical adder [13, 14], these systems are complex and most of them cannot be reduced in size. Therefore, the search of new materials and techniques has become a challenge, especially in applications where the use of ring resonator system and dark-bright soliton conversion behaviors are recommended to overcome the previous problems. In this chapter, we propose an all-optical circuit for logical operations,

Nanophotonics: Devices, Circuits, and Systems
Preecha P. Yupapin, Keerayoot Srinuanjan, and Surachart Kamoldilok
Copyright © 2013 Pan Stanford Publishing Pte. Ltd.
ISBN 978-981-4364-36-2 (Hardcover), 978-981-4364-37-9 (eBook)
www.panstanford.com

which can be used as an alternative to an electronic circuit [15–18]. The theoretical background of an all-optical circuit is also reviewed in this chapter. In principle, the simultaneous addition and subtraction operation of binary based on dark-bright soliton conversion behaviors can be performed, in which the coincident dark and bright soliton pulses are separated after propagating them into a $\pi/2$ phase-shifted device (an optical coupler) [19]. The proposed scheme is based on a 1-bit binary compared to the complex logic circuits, which can be compared to any 2-bit binary, where logics "0" and "1" use the dark and bright soliton pulses, respectively.

3.2 Operating Principle

To begin the all-optical adder/subtractor operation, we consider an electronic circuit that uses the half adder/subtractor shown in Fig. 3.1. In this circuit, the dark-bright soliton conversion pulse [19–22] has been separated into two sets of coupled waveguides by using an optical channel dropping filter (OCDF) [23–27], as shown in Figs. 3.2a, b,

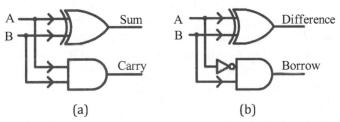

(a)   (b)

Figure 3.1   A schematic diagram of (a) an adder (sum) circuit and (b) a subtractor (difference) circuit.

The relative phase of the two output light signals, after coupling into the optical coupler and before coupling into the ring and the input bus, is $\pi/2$. This means that the signals that couple into the drop and through ports acquire a phase of π with respect to the input port signal.

If the coupling coefficients are formed appropriately, the field coupling into the through port would completely extinguish the resonant wavelength and all power would couple into the drop port, in which the dark-bright conversion behaviors described by Eqs. 3.1–3.8, take place.

$$E_{ra} = -j\kappa_1 E_i + \tau_1 E_{rd} \tag{3.1}$$

$$E_{rb} = \exp(j\omega T/2)\exp(-\alpha L/4)E_{ra} \tag{3.2}$$

$$E_{rc} = \tau_2 E_{rb} - j\kappa_2 E_a \tag{3.3}$$

$$E_{rd} = \exp(j\omega T/2)\exp(-\alpha L/4)E_{rc} \tag{3.4}$$

$$E_t = \tau_1 E_i - j\kappa_1 E_{rd} \tag{3.5}$$

$$E_d = \tau_2 E_a - j\kappa_2 E_{rb} \tag{3.6}$$

Here E_i is the input field, E_a the added (control) field, E_t the throughput field, E_d the dropped field, $E_{ra}...E_{rd}$ the fields in the ring at points $a...d$, respectively, κ_1 the field coupling coefficient between the input bus and the ring, κ_2 the field coupling coefficient between the ring and the output bus, L the circumference ($2\pi R$) of the ring, T the time taken for one round-trip ($T = Ln_{eff}/c$), and α the power loss in the ring per unit length. We assume that lossless coupling, is $\tau_{1,2} = \sqrt{1 - \kappa_{1,2}^2}$.

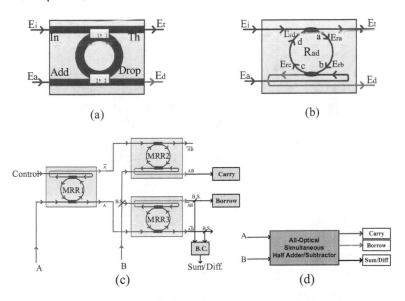

(a) (b) (c) (d)

Figure 3.2 A Schematic diagram of an all-optical half adder/subtractor, in which (a) add/drop filter, (b) modified add/drop filter, and (c) and (d) are half adder and subtractor circuits.

The output power/intensities at the drop port and the through port are given by

$$|E_d|^2 = \left| \frac{-\kappa_1 \kappa_2 A_{1/2} \Phi_{1/2}}{1 - \tau_1 \tau_2 A\Phi} E_i + \frac{\tau_2 - \tau_1 A\Phi}{1 - \tau_1 \tau_2 A\Phi} E_a \right|^2 \tag{3.7}$$

$$|E_t|^2 = \left| \frac{\tau_2 - \tau_1 A\Phi}{1 - \tau_1 \tau_2 A\Phi} E_i + \frac{-\kappa_1 \kappa_2 A_{1/2} \Phi_{1/2}}{1 - \tau_1 \tau_2 A\Phi} E_a \right|^2 \tag{3.8}$$

Here $A_{1/2} = \exp(-\alpha L/4)$ (the half round-trip amplitude), $A = A_{1/2}^2$, $\Phi_{1/2} = \exp(j\omega T/2)$ (the half round-trip phase contribution), and $\Phi = \Phi_{1/2}^2$. The input and control fields at the input and add ports are formed by the dark and bright optical soliton pulses as shown in Eqs. 3.9 and 3.10 [19].

$$E_{in}(t) = A_0 \tanh\left[\frac{T}{T_0}\right] \exp\left[\left(\frac{z}{2L_D}\right) - i\omega_0 t\right] \tag{3.9}$$

$$E_{in}(t) = A_0 \mathrm{sech}\left[\frac{T}{T_0}\right] \exp\left[\left(\frac{z}{2L_D}\right) - i\omega_0 t\right] \tag{3.10}$$

Here A and z are optical field amplitude and propagation distance, respectively. T is the soliton pulse propagation time in a frame moving at the group velocity $T = t - \beta_1 z$, where β_1 and β_2 are the coefficients of the linear and second-order terms of Taylor expansion of the propagation constant. $L_D = T_0^2/|\beta_2|$ is the dispersion length of the soliton pulse. T_0 is the initial soliton pulse width, where t is the soliton phase-shift time and ω_0 the frequency shift of the soliton. This solution describes a pulse that keeps its temporal width invariance as it propagates and is, thus, called a temporal soliton. When the soliton peak intensity $(\beta/\Gamma T_0^2)$ is given, then T_0 is known. For a soliton pulse in a nanoring device, balance should be achieved between the dispersion length (L_D) and nonlinear length $L_{NL} = (1/\Gamma\phi_N)$, where $\Gamma = n_2 k_0$ is the scale over which the dispersive or nonlinear effects make the beam become wider or narrower. For a soliton pulse having balance light properties between dispersion and nonlinear lengths, $L_D = L_{NL}$.

When light propagates within a nonlinear material (medium), the refractive index (n) of the light within the medium is given by Eq. 3.11.

$$n = n_0 + n_2 I = n_0 + \frac{n_2}{A_{eff}} P \tag{3.11}$$

Here n_0 and n_2 are the linear and nonlinear refractive indexes, respectively. I and P are the optical intensity and optical power, respectively. The effective mode core area of the device is given by A_{eff}. For a micro/nanoring resonator, the effective mode core areas range from 0.10 μm² to 0.50 μm² [28]. The resonant output of the light field is the ratio between the output and input fields, $E_{out}(t)$ and $E_{in}(t)$, in each round-trip [21, 29].

3.3   Simultaneous Half Adder/Subtractor

Binary arithmetic is performed like decimal arithmetic, which is presented by the logic gate operation. In the design of an all-optical adder/subtractor circuit, for simplicity, the multiple input ports are required to perform the operation, first consider the half adder/subtractor truth table (Table 3.1). For the half adder/subtractor with two binary inputs, a simplified Boolean equation is obtained and in which the sum of product for each output is given by Eqs. 3.12–3.15 for the half adder and half subtractor systems.

Table 3.1   Truth table of the half adder/subtractor

Input		Half adder		Half subtractor	
A	B	Sum	Carry	Diff.	Borrow
0	0	0	0	0	0
0	1	1	0	1	1
1	0	1	0	1	0
1	1	0	1	0	0

The simplified output of the sum and difference can also be implemented with the XOR gate, in which the addition and subtraction operations can be combined into one circuit with one common binary adder.

$$\text{Sum} = A \oplus B \tag{3.12}$$

$$\text{Carry} = AB \tag{3.13}$$

$$\text{Difference} = A \oplus B \tag{3.14}$$

$$\text{Borrow} = \overline{A}B \tag{3.15}$$

An all-optical half adder/subtractor system is shown in Fig. 3.2c. When the input and control light pulse trains are input into the first add/drop optical filter (MRR1), the dark soliton pulses (logic "0") or the bright soliton pulses (logic "1") are formed within the device. First, the dark soliton pulse is converted to dark-bright soliton via the add/drop optical filter [30], which can be seen at the through and drop ports with π phase shift [31], which, in turn, forms the inverter gate (NOT gate). By using the dark-bright soliton conversion behavior and the add/drop optical filter (MRR2 and MRR3), both input signals are generated again by the first-stage add/drop optical filter. In the next step, the input data "B" with logic "0" (dark soliton) and logic "1" (bright soliton) are added into both add ports, then the dark-bright soliton conversion behavior, with the phase shift π, is operated again. For large-scale modeling (Fig. 3.2c), the optical logic operation results obtained are simultaneously seen at the drop ports (D2 and D3) and through ports (T2 and T3) of the microring resonator MRR2 and MRR3 respectively.

In Fig. 3.2c, the optical logic operation using dark-bright soliton conversion behavior can be described as follows. When input and add ports feed an optical pulse train A, B each into MRR2, the optical pulse trains that appear at the through and drop ports of MRR2 are $\overline{A} \cdot B$ and $A \cdot B$, respectively, whereas the aforementioned assumption is provided. Here, the symbol represents the logic operation AND. Similarly, when input and add ports each feed an optical pulse train A, B into MRR3, the optical pulse trains that appear at the through and drop ports of MRR3 are $\overline{A} \cdot B$ and $A \cdot \overline{B}$, respectively. The all-optical half adder/subtractor can be easily generated by using a beam splitter (BS) or a beam combiner (BC) [8]. The beam splitter to be used should not be a polarizing one and must reflect (and transmit) 50% of the light that is incident as input power.

In simulation, the add/drop optical filter [28, 32] parameters are used and fixed as κ_s = 0.5, R_{ad} = 3.0 mm [31], A_{eff} = 0.25 μm^2, α = 0.05 dBmm^{-1}, n_{eff} = 3.14 (for InGaAsP/InP), and γ = 0.01 for all add/drop optical filters in the system. Results of the simultaneous half adder/subtractor are generated by using the dark-bright soliton conversion behaviors, with wavelength center at λ_0 = 1.50 μm pulse width of 35 fs, and the input data logic "0" and "1" represented by the dark and bright soliton pulses, respectively. In Fig. 3.3, the simultaneous output optical logic gate is seen, which can be configured as follows:

Case 1: When the simultaneous output logic gate input data logic "**00**" is added, the obtained output optical logic is "**1000**" (see Fig. 3.3a).

Case 2: When the simultaneous output logic gate input data logic "**01**" is added, the output optical logic "**0001**" is formed (see Fig. 3.3b).

Case 3: When the simultaneous output logic gate input is "**10**" added, the output optical logic "**0010**" is formed (see Fig. 3.3c).

Case 4: When the simultaneous output logic gate input data logic "**11**" is added, the output optical logic "**0100**" is obtained (see Fig. 3.3d).

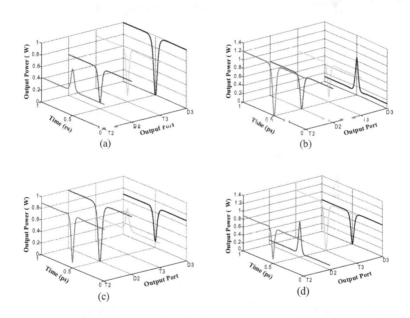

(a)

(b)

(c)

(d)

Figure 3.3 Simulation results of the output logic gates when the input logic states are (a) "DD," (b) "DB," (c) "BD," and (d) "BB."

The simultaneous all-optical half adder/subtractor is shown in Table 3.2. We found that for the output logic in the drop port, T2 and D2 are the optical logic XNOR gates, whereas for the output logic in the through ports, T3 and D3 are the optical logic XOR gates, where logic gate operations can be formed for half adder/subtractor arithmetic simultaneously (see Fig. 3.2c).

Table 3.2 Conclusion output of an optical logic gate

A	B	(T2) $\overline{A} \cdot B$	(D2) $A \cdot B$	(T3) $A \cdot \overline{B}$	(D3) $\overline{A} \cdot B$	Sum/Diff. (XOR) $A \cdot \overline{B} + \overline{A} \cdot B$
D	D	B	D	D	D	D
D	B	D	D	D	B	B
B	D	D	D	B	D	B
B	B	D	B	D	D	D

Note: D refers to dark soliton = logic "0" and B refers to bright soliton = logic "1."

3.4 Simultaneous Full Adder/Subtractor

In operation, a full adder/subtractor is a combined circuit that forms the arithmetic of three bits input and two outputs. The binary signals of the inputs are considered to be binary digits and, when operated arithmetically, form simultaneous sum and difference at the output (the through and drop ports). The binary values are considered as variables of Boolean function for a circuit implemented with logic gates (Table 3.3). Equations 3.16–3.19 give the sum of the products of each output of the full adder/subtractor.

$$\text{Sum} = A \oplus B \oplus C_i \tag{3.16}$$

$$\text{Carry} = AB + C_i \, (A \oplus B) \tag{3.17}$$

$$\text{Difference} = A \oplus B \oplus B_i \tag{3.18}$$

$$\text{Borrow} = \overline{A}B + B_i \, (\overline{A \oplus B}) \tag{3.19}$$

Table 3.3 Truth table of a full adder/subtractor

Input			Full adder		Full subtractor	
A	B	C_i/B_i	Sum	Carry	Diff.	Borrow
0	0	0	0	0	0	0
0	0	1	1	0	1	1
0	1	0	1	0	1	0
0	1	1	0	1	0	0
1	0	0	1	0	1	1
1	0	1	0	1	0	1
1	1	0	0	1	0	0
1	1	1	1	1	1	1

Figure 3.4 shows the full adder and full subtractor electronic circuits while Fig. 3.5 shows a simultaneous full adder/subtractor using an optical system, which is a modified simultaneous half adder/subtractor shown in Fig. 3.2. In Fig. 3.5, the input "A" and control light pulse trains are added into the first add/drop optical filter (MRR1) in the first stage, where the dark-bright soliton conversion behavior is seen at the through and drop ports with π phase shift. In the second stage (MRR2, MRR3), both inputs are regenerated by the add/drop optical filter in the first stage, where the input data "B" with logic "0" (dark soliton) and logic "1" (bright soliton) are added to both add ports. The second stage outputs are dark-bright soliton conversion pulses again with π phase shift. The third-stage operation is formed by the T3 and D3 outputs from the second stage of the add/drop optical filter (MRR4), where the dark-bright soliton conversion behavior is seen at the through and drop ports with π phase shift. In the fourth stage (MRR5, MRR6), both inputs are regenerated by the third stage of the add/drop optical filter, where the input data "C" with logic "0" (dark soliton) and logic "1" (bright soliton) are added to both add ports. The outputs of the fourth stage are dark-bright soliton conversion behavior again with π phase shift. In Fig. 3.5, the simultaneous output optical logic gate is seen.

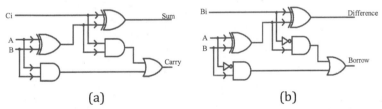

Figure 3.4   A schematic diagram of (a) full adder and (b) full subtractor circuits.

In Fig. 3.6, the simultaneous output optical logic gate is seen, which can be configured as follows:

Case 1: When the simultaneous output logic gate input data logic "**000**" is added, the obtained output optical logic is "**10001000**" (see Fig. 3.6a).

Case 2: When the simultaneous output logic gate input data logic "**001**" is added, the output optical logic "**10000001**" is formed (see Fig. 3.6b).

Case 3: When the simultaneous output logic gate input "**010**" is added, the output optical logic "**00010010**" is formed (see Fig. 3.6c).

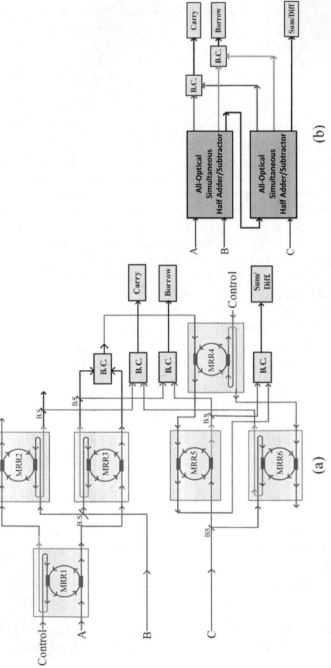

Figure 3.5 A Schematic diagram of a all-optical full adder/subtractor, in which (a) a full adder circuit and (b) a full subtractor circuit.

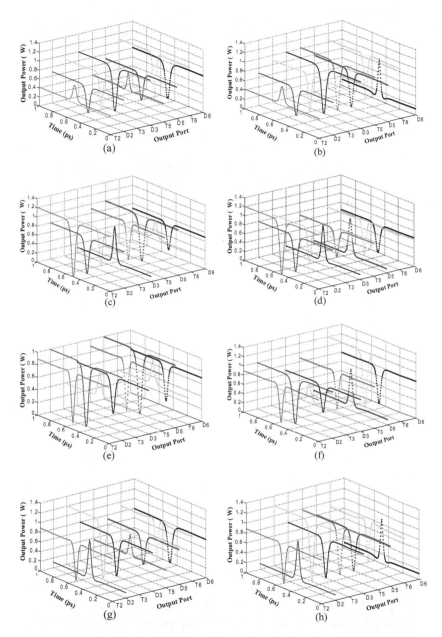

Figure 3.6 Simulation results of the output logic when the input logic states are (a) "DDD," (b) "DDB," (c) "DBD," (d) "DBB," (e) "BDD," (f) "BDB," (g) "BBD," and (h) "BBB."

Case 4: When the simultaneous output logic gate input data logic "**011**" is added, the output optical logic "**00010100**" is obtained (see Fig. 3.6d).

Case 5: When the simultaneous output logic gate input data logic "**100**" is added, the obtained output optical logic is "**00100010**" (see Fig. 3.6e).

Case 6: When the simultaneous output logic gate input data logic "**101**" is added, the output optical logic "**00100100**" is formed (see Fig. 3.6f).

Case 7: When the simultaneous output logic gate input data logic "**110**" is added, the output optical logic "**01001000**" is formed (see Fig. 3.6g).

Case 8: When the simultaneous output logic gate input data logic "**111**" is added, the output optical logic "**01000001**" is obtained (see Fig. 3.6h).

The simulation results of the simultaneous full adder/subtractor outputs are obtained when the input data "ABC" are (a) "DDD," (b) "DDB," (c) "DBD," (d) "DBB," (e) "BDD," (f) "BDB," (g) "BBD," and (h) "BBB". Results of all outputs are shown in Table 3.4.

Table 3.4 Conclusion output of the optical logic gate

A	B	C	(T2)	(D2)	(T3)	(D3)	(T5)	(D5)	(T6)	(D6)
D	D	D	B	D	D	D	B	D	D	D
D	D	B	B	D	D	D	D	D	D	B
D	B	D	D	D	D	B	D	D	B	D
D	B	B	D	D	D	B	D	B	D	D
B	D	D	D	D	B	D	D	D	B	D
B	D	B	D	D	B	D	D	B	D	D
B	B	D	D	B	D	D	B	D	D	D
B	B	B	D	B	D	D	D	D	D	B

Note: D refers to the dark soliton = logic "0" and B refers to bright soliton = logic "1."

3.5 Conclusion

The operation of simultaneous addition/subtraction logic has been proposed by using dark-bright soliton conversion behavior system via the modified add/drop filters. By using the dark-bright soliton conversion behavior concept, the data logic "0" (dark soliton) and "1" (bright soliton), which is all-optical in nature, can be used to form the simultaneous addition/subtraction logic operations in which the logic status results can be obtained simultaneously at the drop and through ports. Therefore, the proposed design can be used in a logical circuit, which is recognized as a simple and flexible system for performing logic switching. Moreover, such devices can be improvised to be used for any number of input digits by the proper incorporation of optical switches based on dark-bright soliton conversion behavior and, therefore, can be used in advanced applications.

References

1. J. H. Kim, Y. T. Byun, Y. M. Jhon, S. Lee, D. H. Woo, and S. H. Kim, "All-optical half adder using semiconductor optical amplifier based devices," *Opt. Commun.*, **218**, 345–349 (2003).

2. J. Cong, X. Zhang, and D. Huang, "A propose for two-input arbitrary Boolean logic gates using single semiconductor optical amplifier by picosecond pulse injection," *Opt. Express*, **17**, 7725–7730 (2009).

3. S. H. Kim, J. H. Kim, B. G. Yu, Y. T. Byun, Y. M. Jeon, S. Lee, and D. H. Woo, "All-optical NAND gate using cross-gain modulation in semiconductor optical amplifiers," *Electron. Lett.*, **41**, 1027–1028 (2005).

4. J. Hun, Y. M. John, Y. T. Byun, S. Lee, D. H. Woo, and S. H. Kim, "All-optical XOR gate using semiconductor optical amplifiers without additional input beam," *IEEE Photon. Technol. Lett.*, **14**, 1436–1438 (2002).

5. T. Kawazoe, K. Kobayashi, K. Akahane, M. Naruse, N. Yamamoto, and M. Ohtsu, "Demonstration of nanophotonic NOT gate using near-field optically coupled quantum dot," *Appl. Phys. B*, **84**, 243–246 (2006).

6. S. Ma, Z. Chen, H. Sun, and K. Dutta, "High speed all optical logic gates based on quantum dot semiconductor optical amplifiers," *Opt. Express*, **18**, 6417–6422 (2010).

7. A. J. Poustie, K. J. Blow, R. J. Manning, and A. E. Kelly, "All-optical pseudorandom number generator," *Opt. Commun.*, **159**, 208–214 (1999).

8. J. N. Roy, and D. K. Gayen, "Integrated all-optical logic and arithmetic operation with the help of a TOAD-based interferometer device-alternative approach," *Appl. Opt.*, **46**, 5304–5310 (2007).

9. L. Zhang, R. Ji, L. Jia, L. Yang, P. Zhou, Y. Tiam, P. Chen, and Y. Lu, "Demonstration of directed XOR/XNOR logic gates using two cascaded microring resonators," *Opt. Lett.*, **35**(10), 1620–1622 (2010).

10. D. K. Gayen, and J. N. Roy, "All-optical arithmetic unit with the help of terahertz-optical-asymmetric-demultiplexer-based tree architecture," *Appl. Opt.*, **47**, 933–943 (2008).

11. J. N. Roya., A. K. Maitib, D. Samantac, and S. Mukhopadhyayc, "Tree-net architecture for integrated all-optical arithmetic operations and data comparison scheme with optical nonlinear material," *Opt. Switch. Netw.*, **4**, 231–237 (2007).

12. A. Poustie, R. Manning, A. Kelly, and K. Blow, "All-optical binary counter," *Opt. Express*, **6**, 69–74 (2000).

13. N. Pahari, D. N. Das, and S. Mukhopadhyay, "All-optical method for the addition of binary data by nonlinear materials," *Appl. Opt.*, **43**(33), 6147–6150 (2004).

14. A. J. Poustie, K. J. Blow, A. E. Kelly, and R. J. Manning, "All-optical full adder with bit-differential delay," *Opt. Commun.*, **168**, 89–93 (1999).

15. S. Mukhopadhyay, J. N. Roy, and S. K. Bera, "Design of minimized LED array for maximum parallel logic operations in optical shadow casting technique," *Opt. Commun.*, **99**, 31–37 (1993).

16. D. E. Rumelhart, and J. L. McClelland (eds), "Parallel distributed processing: Explorations in the microstructure of cognition," Vols. 1 and 2, MIT Press, Boston, MA (1986).

17. N. Peyghambarian, and H. M. Gibbs, "Optical bistability for optical signal processing and computing," *Opt. Eng.*, **24**(1), 68–72 (1985).

18. N. Pahari, D. N. Das, and S. Mukhopadhyay, "All-optical method for the addition of binary data by nonlinear materials," *Appl. Opt.*, **43**(33), 6147–6150 (2004).

19. K. Sarapat, N. Sangwara, K. Srinuanjan, and P. P. Yupapin, "Novel dark-bright optical solitons conversion system and power amplification," *Opt. Eng.*, **48**(4), 045004-7 (2009).

20. S. Mitatha, "Dark soliton behaviors within the nonlinear micro and nanoring resonators and applications," *Prog. Electromagn. Res.*, **99**, 383–404 (2009).

21. T. Phatharaworamet, C. Teeka, R. Jomtarak, S. Mitatha, and P. P. Yupapin, "Random binary code generation using dark-bright soliton conversion control within a PANDA ring resonator," *J. Lightwave Technol.*, **28**(19), 2804–2809 (2010).

22. P. Juleang, P. Phongsanam, S. Mitatha, and P. P. Yupapin, "Public key suppression and recovery using a PANDA ring resonator for high security communication," *Opt. Eng.*, **52**(3), 035002-1-6 (2011).

23. P. P. Absil, J. V. Hryniewicz, B. E. Little, F. G. Johnson, and P. T. Ho, "Vertically coupled microring resonators using polymer wafer bonding," *IEEE Photon. Technol. Lett.*, **13**, 49–51 (2001).

24. R. Grover, P. P. Absil, V. Van, J. V. Hryniewicz, B. E. Little, O. S. King, L. C. Calhoun, F. G. Johnson, and P. T. Ho, "Vertically coupled GaInAsP-InP microring resonators," *Opt. Lett.*, **26**, 506–508 (2001).

25. V. Van, T. A. Ibrahim, P. P. Absil, F. G. Johnson, and R. Grover, "Optical signal processing using nonlinear semiconductor micro ring resonators," *IEEE J. Sel. Top. Quantum Electron.*, **8**, 705–713 (2002).

26. S. Mitatha, N. Pornsuwancharoen, and P. P. Yupapin, "A simultaneous short-wave and millimeter-wave generation using a soliton pulse within a nano-waveguide," *IEEE Photon. Technol. Lett.*, **13**, 932–934 (2009).

27. P. P. Yupapin, and N. Pornsuwancharoen, "Guided wave optics and photonics: Micro ring resonator design for telephone network security," Nova Science Publishers, New York (2008).

28. Q. Xu, D. Fattal, and R. G. Beausoleil, "Silicon microring resonators with 1.5-μm radius," *Opt. Express*, **16**, 4309–4315 (2008).

29. P. P. Yupapin, N. Pornsuwanchroen, and S. Chaiyasoonthorn, "Attosecond pulse generation using nonlinear micro ring resonator," *Microw. Opt. Technol. Lett.*, **50**, 3108–3011 (2008).

30. S. Mitatha, N. Chaiyasoonthorn, and P. P. Yupapin, "Dark-bright optical solitons conversion via an optical add/drop filter," *Microw. Opt. Technol. Lett.*, **51**, 2104–2107 (2009).

31. J. Wang, Q. Sun, and J. Sun, "All-optical 40 Gbit/s CSRZ-DPSK logic XOR gate and format conversion using four-wave mixing," *Opt. Express*, **17**(15), 12555–12563 (2009).

32. S. Mookherjea, and M. A. Schneider, "The nonlinear microring add-drop filter," *Opt. Express*, **16**, 15130–15136 (2008).

Chapter 4

Photonic Transistor Manipulation

4.1 Introduction

Optical transistors have been widely investigated in recent years [1–8], and they have got substantial publicity since then. From the term "optical transistor," one can guess that it will be some kind of photonic analog of an electronic transistor that can perform similar functions. An optical transistor is considered important because it allows the construction of sophisticated optical circuits. Like the electronic transistor, amplification is an important feature of an optical transistor. It is simply not sufficient to influence one light beam with another beam. Recently, Yupapin *et al.* [9–15] have presented an all-optical device that consists of an add/drop optical filter. The device, known as a PANDA ring resonator, was constructed using optical dark-bright soliton control within a semiconductor add/drop multiplexer and has shown promising results for future applications. The use of this device in various applications has been investigated thoroughly by scientists [16, 17]. One of the advantages of the PANDA ring resonator is that the dark soliton peak signal is always at a low level, which is useful for secured signal communication in a transmission link. The other dark soliton peak signal is formed when a high optical field is configured as an optical tweezer or a potential well [18, 19]. In this chapter, a novel type of transistor, known as a hybrid transistor, is proposed. The hybrid transistor can be changed to all types of transistors by using the PANDA ring resonator. In principle, the various transistors made from the hybrid transistor are

Nanophotonics: Devices, Circuits, and Systems
Preecha P. Yupapin, Keerayoot Srinuanjan, and Surachart Kamoldilok
Copyright © 2013 Pan Stanford Publishing Pte. Ltd.
ISBN 978-981-4364-36-2 (Hardcover), 978-981-4364-37-9 (eBook)
www.panstanford.com

formed by using the atom/molecule trapping tool called a tweezer. All types of transistors that are formed are also known as hybrid transistors, which will be described in detail in the last section of this chapter.

4.2 Operating Principle

An optical transistor is a novel system of photonic transistor in which a PANDA microring resonator is used. In this device, dark-bright soliton pulses are propagated within an add/drop optical multiplexer of the PANDA microring. The multiplexed signals with slightly different wavelengths of the dark solitons are controlled and amplified within the system. In this chapter, the dynamic behaviors of dark-bright soliton interactions have been analyzed and described. Finally, the use of optical switching to form a hybrid photonic transistor by using Gaussian control at the add port is discussed in detail.

A stationary dark soliton pulse is introduced in the add/drop optical filter system shown in Fig. 4.1. The input optical field (E_{in}) and the add port optical field (E_{add}) of the dark-bright solitons or Gaussian pulses are given by the following equations [20]:

$$E_{in}(t) = A\tanh\left[\frac{T}{T_0}\right]\exp\left[\left(\frac{z}{2L_D}\right) - i\omega_0 t\right]$$ (4.1a)

$$E_{add}(t) = A\,\mathrm{sech}\left[\frac{T}{T_0}\right]\exp\left[\left(\frac{z}{2L_D}\right) - i\omega_0 t\right]$$ (4.1b)

$$E_{add}(t) = E_0\exp\left[\left(\frac{z}{2L_D}\right) - i\omega_0 t\right]$$ (4.1c)

Here A and x are the optical field amplitude and propagation distance, respectively. $T = t - \beta_1 x$, where T is the soliton pulse propagation time in a frame moving at the group velocity, and $L_D = T_0^2/|\beta_2|$ is the dispersion length of the soliton pulse. β_1 and β_2 are the coefficients of the linear and second-order terms of Taylor expansion of the propagation constant. T_0 is the soliton pulse propagation time at initial input (or the soliton pulse width), t the soliton phase-shift time, and ω_0 the frequency shift of the soliton. This solution describes a pulse that keeps its temporal width invariance as it propagates and is, therefore, called a temporal soliton. When a soliton of peak intensity $(|\beta_2/\Gamma T_0^2|)$ is given, then T_0 is known. For a soliton pulse in a microring device, a balance should be achieved between the dispersion length (L_D) and the nonlinear length $(L_{NL} = 1/\Gamma\phi_{NL})$, where $\Gamma = n_2 k_n$ is the length scale over which dispersive or nonlinear effects make the beam wider or narrower. For a soliton pulse, there is a balance between dispersion and nonlinear lengths. Hence $L_D = L_{NL}$. For a Gaussian pulse (Eq. 4.1c), E_0 is the amplitude of the optical field.

When light propagates within a nonlinear medium, the refractive index (n) of light within the medium is given by

$$n = n_0 + n_2 I = n_0 + \frac{n_2}{A_{eff}} P \qquad (4.2)$$

Here n_0 and n_2 are the linear and nonlinear refractive indexes, respectively. I and P are the optical intensity and power, respectively. The effective mode core area of the device is given by A_{eff}. For the add/drop optical filter design, the effective mode core areas range from 0.10 μm² to 0.50 μm² and the parameters were obtained by using the related practical material parameters (InGaAsP/InP) [21]. When a dark soliton pulse is input and propagated within an add/drop optical filter as shown in Fig. 4.1, a resonant output is formed.

The resonator output fields, E_{t1} and E_1, consist of transmitted and circulated components of the add/drop optical filter system, which can act as driving forces for a photon/molecule/atom.

For the first coupler of the add/drop optical filter system, the transmitted and circulated components can be written as

$$E_{t1} = \sqrt{1-\gamma_1}\left[\sqrt{1-\kappa_1}\,E_{i1} + j\sqrt{\kappa_1}\,E_4\right] \qquad (4.3)$$

$$E_1 = \sqrt{1-\gamma_1}\left[\sqrt{1-\kappa_1}\,E_4 + j\sqrt{\kappa_1}\,E_{i1}\right] \qquad (4.4)$$

$$E_2 = E_0 E_1 e^{-\frac{\alpha L}{2}\frac{1}{2} - jk_n\frac{L}{2}} \qquad (4.5)$$

Here κ_1 is the intensity coupling coefficient, γ_1 the fractional coupler-intensity loss, α the attenuation coefficient, $k_n = 2\pi/\lambda$ the wave propagation number, λ the input wavelength light field, and $L = 2\pi R_{ad}$, where R_{ad} is the radius of the add/drop device.

For the second coupler of the add/drop system,

$$E_{t2} = \sqrt{1-\gamma_2}\left[\sqrt{1-\kappa_2}E_{i2} + j\sqrt{\kappa_2}E_2\right] \tag{4.6}$$

$$E_3 = \sqrt{1-\gamma_2}\left[\sqrt{1-\kappa_2}E_2 + j\sqrt{\kappa_2}E_{i2}\right] \tag{4.7}$$

$$E_4 = E_{0L}E_3 e^{-\frac{\alpha L}{2}\frac{L}{2} - jk_n\frac{L}{2}} \tag{4.8}$$

Here κ_2 is the intensity coupling coefficient, γ_2 the fractional coupler-intensity loss. The circulated light fields, E_0 and E_{0L}, the light field circulated components of the nanorings of radii R_r and R_L, which couple into the right and left sides of the add/drop optical filter system, respectively. The light field transmitted and circulated components in the right nanoring are given by

$$E_2 = \sqrt{1-\gamma}\left[\sqrt{1-\kappa_0}E_1 + j\sqrt{\kappa_0}E_{r2}\right] \tag{4.9}$$

$$E_{r1} = \sqrt{1-\gamma}\left[\sqrt{1-\kappa_0}E_{r2} + j\sqrt{\kappa_0}E_1\right] \tag{4.10}$$

$$E_{r2} = E_{r1}e^{-\frac{\alpha}{2}L_1 - jk_n L_1} \tag{4.11}$$

Here κ_0 is the intensity coupling coefficient, γ the fractional coupler intensity loss, α the attenuation coefficient, $k_n = 2\pi/\lambda$ the wave propagation number, λ the input wavelength light field, and $L_1 = 2\pi R_r$, where R_r is the radius of right nanoring.

The output optical field of the through port (E_{t1}) is expressed by

$$E_{t1} = AE_{i1} - BE_{i2}e^{-\frac{\alpha L}{2}\frac{L}{2} - jk_n\frac{L}{2}}\left[\frac{CE_{i1}e^{-\frac{\alpha}{2}L - jk_n L} + DE_{i2}e^{-\frac{3\alpha L}{2}\frac{L}{2} - jk_n\frac{3L}{2}}}{1 - Fe^{-\frac{\alpha}{2}L - jk_n L}}\right], \tag{4.12}$$

Here $A = x_1 x_2$, $B = x_1 x_2 y_2\sqrt{\kappa_1}E_{0L}$, $C = x_1^2 x_2 \kappa_1\sqrt{\kappa_2}E_0 E_{0L}$, $D = (x_1 x_2)^2 y_1 y_2 \sqrt{\kappa_1 \kappa_2}E_0 E_{0L}^2$, and $F = x_1 x_2 y_1 y_2 E_0 E_{0L}$.

The power output of the through port (P_{t1}) is given as

$$P_{t1} = (E_{t1})\cdot(E_{t1})^* = |E_{t1}|^2 . \tag{4.13}$$

Similarly, the output optical field of the drop port (E_{t2}) is given as

$$E_{t2} = x_2 y_2 E_{i2} - \left[\frac{x_1 x_2\sqrt{\kappa_1 \kappa_2}E_0 E_{i1}e^{-\frac{\alpha l}{2}\frac{L}{2} - jk_n\frac{L}{2}} + x_1 x_2^2 y_1 y_2\sqrt{\kappa_2}E_0 E_{0L}e^{-\frac{\alpha}{2}l - jk_n L}}{1 - x_1 x_2 y_1 y_2 E_0 E_{0L}e^{-\frac{\alpha}{2}l - jk_n L}}\right] \tag{4.14}$$

The power output of the drop port (P_{t2}) is expressed as

$$P_{t2} = (E_{t2}) \cdot (E_{t2})^* = |E_{t2}|^2 .$$ (4.15)

In order to retrieve the required amplification and optical switching signals of a photonic transistor, we propose to use an add/drop device with appropriate parameters, the details of which are given in the following text. The optical circuits of ring resonator add/drop filters for the through port and drop port can be given by Eqs. 4.13 and 4.15, respectively. κ_1 and κ_2 are the coupling coefficients of the add/drop filters, $k_n = 2\pi/\lambda$ the wave propagation number in vacuum, and $\alpha = 5 \times 10^{-5}$ dBmm^{-1} the waveguide (microring resonator) loss. The fractional coupler-intensity loss is $\gamma = 0.01$. In the case of an add/drop device, the nonlinear refractive index is neglected.

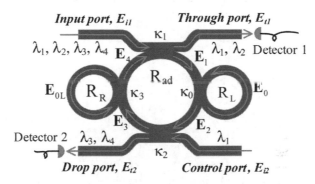

Figure 4.1   A schematic diagram of light signal multiplexer controlled by light.

4.3   Photonic Transistor Characteristics

To manipulate the control multiplexer operation, simulation results of the dynamic optical amplification and switching signals within the light signal multiplexer should be as shown in Fig. 4.2. In this case, the dark soliton is input into the add port and the dynamic optical photonic transistor output signals shown in Figs. 4.2a–f are both amplification and switching characteristics. When the output power, controlled by light at the control port, was increased to 800 W and above at the through port (Fig. 4.2e), the output power is dropped to 9.8 W at the drop port, which can be used to trap the photons/atoms/molecules trapping (Fig. 4.2f). Figure 4.3 shows how the optical switching power signal multiplexer can be used to form a hybrid photonic transistor and how photons/molecules/atoms can be fed into the light signal multiplexer by a dark soliton. The output photons/molecules/atoms from the through port are made to enter the target and reflect back to the light signal multiplexer. The induced change of the collision (coupling effects) can be controlled by the add port input signal and, finally, the interference signal can be seen at the drop and through ports. This can be done by balancing and adjusting the controlled parameters via the add port (control port). In Figs. 4.4 and 4.5, where plots of output and input signal relationships have been shown, the interesting characteristic is that desired linear gains can be obtained by using different coupling coefficient values. Figure 4.4 shows the variations in the output power for an input power range of 1–15 mW and for fixed values of other parameters as given in the figure caption.

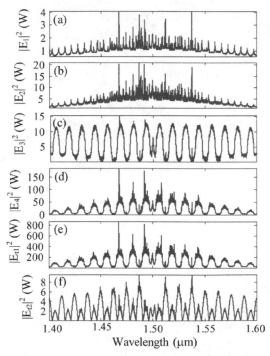

Figure 4.2 Simulation results of the optical amplification and switching signals for photonic transistor with center wavelength 1.50 μm, where (a) $|E_1|^2$, (b) $|E_2|^2$, (c) $|E_3|^2$, (d) $|E_4|^2$ and (e) the through port signal, and (f) the drop port signal.

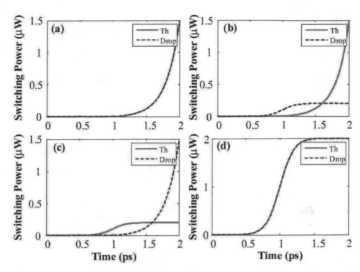

Figure 4.3 Graphs showing the switching power of a photonic transistor using PANDA microring, for which the input and control signals are (a) dark-dark solitons (b) dark-bright solitons (c) bright-dark solitons and (d) bright-bright solitons.

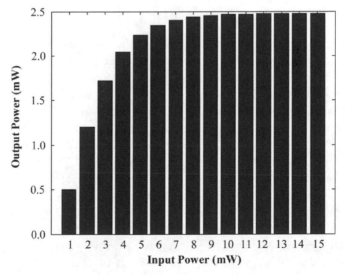

Figure 4.4 Graph showing the switching power of a photonic transistor using PANDA microring, in which the Gaussian input was controlled at the add port and values of the given parameters were fixed as $\kappa = \kappa_1 = \kappa_2 = \kappa_3 = 0.5$, $R_{ad} = 15$ μm, and $R_R = R_L = 3$ μm.

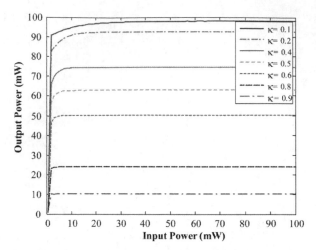

Figure 4.5   Graph showing the switching power of a photonic transistor using PANDA microring, in which the Gaussian input was controlled at the add port and values of the given parameters were fixed as $\kappa = \kappa_2 = \kappa_3 = 0.5$, $R_{ad} = 15$ μm, $R_R = R_L = 3$ μm and κ_1 varied from 0.1 to 0.9.

4.4 Molecular Transistor

By using optical tweezers, which have been proposed recently by Suwanpayak *et al.* [11], the trapped molecules can be used in the same manner as is done in a photonic transistor. By increasing the magnitude of the force, an increase in the number of trapped molecules can be obtained. In Fig. 4.6, the size of the molecular trapping probe can be adjusted from 10 nm to 15 nm according to the size of the atoms/molecules and can be used for atoms/molecules transportation at the through port and storage at the drop port. The number of atoms/molecules can be increased within the PANDA ring resonator as shown in Fig. 4.6a. In addition, the size of the trapping tool (probe) or dynamic well can be adjusted by varying the coupling coefficient of the PANDA microring as shown in Fig. 4.6b, keeping the other parameters fixed at the values given in the figure caption. In Fig. 4.7, when the right and left PANDA microring radii were increased from 3 μm to 6 μm, an increase of about 5 nm in the width of the trapping probe was seen. The number of molecules/atoms can be controlled to achieve the constant and amplification characteristics required for the performance of a molecular transistor.

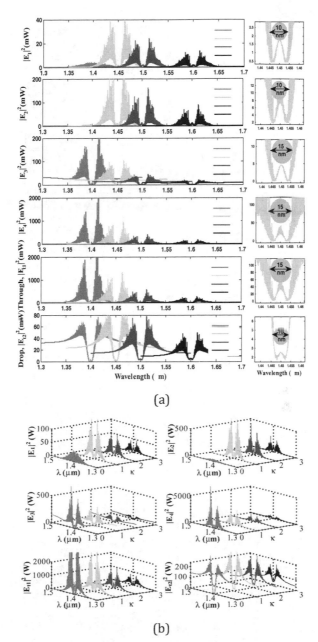

(a)

(b)

Figure 4.6 Results of dynamic tweezers within an optical transistor for different (a) wavelengths and (b) coupling constants, with R_{ad} = 15 μm and R_R = R_L = 3 μm.

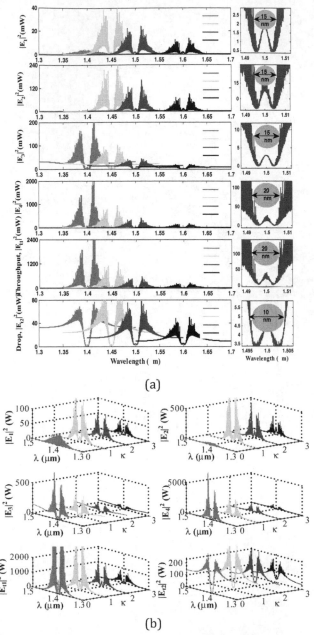

(a)

(b)

Figure 4.7 Results of dynamic tweezers within an optical transistor for different (a) wavelengths and (b) coupling constants, with R_{ad} = 15 μm and $R_R = R_L$ = 6 μm.

4.5 Conclusion

In this chapter, the method of forming a hybrid photonic transistor by using dark-bright solitons conversion was discussed. The transistor can be formed by incorporating the PANDA microring resonator in the proposed light signal multiplexer. The dynamic behavior can be controlled and used to form hybrid devices. This behavior can also be used to confine a light pulse, atom, or molecule to a suitable size, which, in turn, can be employed as optical tweezers. The term "dynamic probing" is becoming a reality for optical tweezers, in which trapped pulses or molecules within a given period of time (memory) is plausible. The proposed concept and system can be applied to form a nanoscale interpretation of a hybrid interferometer. The trapped photon/molecules/atoms can be used to process the nanoscale signal interpretation based on a photonic transistor, where the balancing parameters can be found (measured). Thus, by using dynamic tweezers, a hybrid photonic transistor using photons/atoms/molecules trapping and transportation within the system can be realized, and this can be done for all types of transistors.

References

1. J. Hwang, M. Pototschnig, R. Lettow, G. Zumofen, A. Renn, S. Gotzinger, and V. Sandoghdar, "A single-molecule optical transistor," *Nature*, **460**, 76–80 (2009).
2. F. Y. Hong, and S. J. Xiong, "Single-photon transistor using microtoroidal resonators," *Phys. Rev. A*, **72**, 013812 (2008).
3. D. E. Chang, A. S. Sorensen, E. A. Demler, and M. D. Lukin, "A single-photon transistor using nanoscale surface plasmons," *Nat. Phys.*, **3**, 807–812 (2007).
4. A. Micheli, A. J. Daley, D. Jaksch, and P. Zoller, "Single atom transistor in a 1D optical lattice," *Phys. Rev. Lett.*, **93**(14), 140408 (2004).
5. A. J. Daley, S. R. Clark, D. Jaksch, and P. Zoller, "Numerical analysis of coherent many-body currents in a single atom transistor," *Phys. Rev. A*, **72**, 043618 (2005).
6. M. F. Yanik, S. Fan, M. Soljacic, and J. D. Joannopoulos, "All-optical transistor action with bistable switching in a photonic crystal cross-waveguide geometry," *Opt. Lett.*, **28**(24), 2506–2508 (2003).
7. Y. Huang, and S. T. Ho, "High-speed low-power photonic transistor devices based on optically-controlled gain or absorption to affect optical interference," *Opt. Express*, **16**(21), 16806–16824 (2008).

8. S. Medhekar, and Ram Krishna Sarkar, "All-optical passive transistor," *Opt. Lett.*, **30**, 887–889 (2005).

9. K. Uomwech, K. Sarapat, and P. P. Yupapin, "Dynamic modulated Gaussian pulse propagation within the double PANDA ring resonator system," *Microw. Opt. Technol. Lett.*, **52**(8), 1818–1821 (2010).

10. T. Phatharaworamet, C. Teeka, R. Jomtarak, S. Mitatha, and P. P. Yupapin, "Random binary code generation using dark-bright soliton conversion control within a PANDA ring resonator," *J. Lightwave Technol.*, **28**(19), 2804–2809 (2010).

11. N. Suwanpayak, M. A. Jalil, C. Teeka, J. Ali, and P. P. Yupapin, "Optical vortices generated by a PANDA ring resonator for drug trapping and delivery applications," *Biomed. Opt. Express*, **2**(1), 159–168 (2011).

12. P. P. Yupapin, N. Suwanpayak, B. Jukgoljun, and C. Teeka, "Hybrid transceiver using a PANDA ring resonator for nanocommunication," *Phys. Express*, **1**(1), 1–9 (2011).

13. M. Tasakorn, C. Teeka, R. Jomtarak, and P. P. Yupapin, "Multitweezers generation control within a nanoring resonator system," *Opt. Eng.*, **49**(7), 075002 (2010).

14. B. Jukgoljun, N. Suwanpayak, C. Teeka, and P. P. Yupapin, "Hybrid transceiver and repeater using a PANDA ring resonator for nanocommunication," *Opt. Eng.*, **49**(12), 125003 (2010).

15. P. Youplao, T. Phattaraworamet, S. Mitatha, C. Teeka, and P. P. Yupapin, "Novel optical trapping tool generation and storage controlled by light," *J. Nonlinear Opt. Phys. Mater.*, **19**(2), 371–378 (2010).

16. K. Sarapat, N. Sangwara, K. Srinuanjan, P. P. Yupapin, and N. Pornsuwancharoen, "Novel dark-bright optical solitons conversion system and power amplification," *Opt. Eng.*, **48**, 045004-1-5 (2009).

17. S. Mitatha, N. Chaiyasoonthorn, and P. P. Yupapin, "Dark-bright optical solitons conversion via an optical add/drop filter," *Microw. Opt. Technol. Lett.*, **51**, 2104–2107 (2009).

18. T. Threepak, X. Luangvilay, S. Mitatha, and P. P. Yupapin, "Novel quantum-molecular transporter and networking via a wavelength router," *Microw. Opt. Technol. Lett.*, **52**(6), 1353–1357 (2010).

19. K. Kulsirirat, W. Techithdeera, and P. P. Yupapin, "Dynamic potential well generation and control using double resonators incorporating in an add/drop filter," *Mod. Phys. Lett. B*, **24**(32), 3071–3080 (2010).

20. S. Mitatha, N. Pornsuwancharoen, and P. P. Yupapin, "A simultaneous short-wave and millimeter-wave generation using a soliton pulse within a nano-waveguide," *IEEE Photon. Technol. Lett.*, **21**, 932–934 (2009).

21. Y. Kokubun, Y. Hatakeyama, M. Ogata, S. Suzuki, and N. Zaizen, "Fabrication technologies for vertically coupled microring resonator with multilevel crossing busline and ultracompact-ring radius," *IEEE J. Sel. Top. Quantum Electron.*, **11**, 4–10 (2005).

Chapter 5

Nanoscale Sensing Device Design

5.1   Introduction

The use of microring resonators in various applications has been widely investigated in theory and in experiments [1–3]. One such interesting application is the use of a specific model of a ring resonator, known as a PANDA ring resonator [4], as a nanoscale sensing and measuring device in high-efficiency systems [5, 6]. A PANDA ring resonator is recommended because it fulfills specific requirements of sensing applications, especially in the nanoscale resolution regime. Recently, some authors have shown that a PANDA ring resonator [7, 8] can establish the new concept of dark-bright soliton conversion, wherein by the use of random encoding, optical vortices (tweezers) and optical/quantum gate can be generated.

In this chapter, we present the use of a PANDA ring resonator as a nanoscale force-sensing application, in which the resolution in the range of nN (nano-Newton) can be achieved by measuring the wavelength shift, and the low power consumption due to the low-intensity source is an added advantage. The sensing system functions when one of the ring radius changes due to a load cell or any other physical parameter, which results in a change in the optical path length of light in the same manner as it happens in an interferometer [9, 10], while the other ring radius remains constant (reference signal). The sensing and reference signals in both the stages can be analyzed, simulated, and compared. Simulation results obtained

Nanophotonics: Devices, Circuits, and Systems
Preecha P. Yupapin, Keerayoot Srinuanjan, and Surachart Kamoldilok
Copyright © 2013 Pan Stanford Publishing Pte. Ltd.
ISBN 978-981-4364-36-2 (Hardcover), 978-981-4364-37-9 (eBook)
www.panstanford.com

show that this system can be employed as a sensing instrument of nanoscale resolution. Although, measurement limitation due to wavelength meter resolution and material elongation limit occurs, a measurement resolution of 1 nm is noted in this chapter.

5.2 Principle and Method

To perform simulation sensing, InGaAsP/InP should be used as the microring material, which must have a refractive index (n_0) of 3.34 [11–13]. A schematic diagram of a sensing system using a PANDA ring resonator is shown in Fig. 5.1. The system consists of three microring resonators. The first ring is placed as a reference ring and has a radius $R_1 = 1.550$ µm. The second ring is the sensing ring and its radius R_2 varies from 1.550 µm to 1.558 µm. The third ring is used to form the interference signal between the reference and sensing rings, and its radius R_3 is 3.10 µm. In operation, the change in the sensing ring radius is caused by a change in the shift of signals circulated in the interferometer ring (R_3), in which the interference signal is seen. The change in the optical path length, which is related to the change of the external parameters, can be compared and measured.

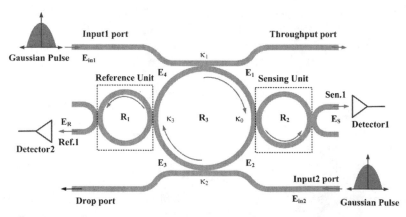

Figure 5.1 A schematic diagram of a nanoscale sensing transducer using PANDA ring resonator.

Two identical beams of monochromatic optical field (E_{in}) of a Gaussian pulse with center wavelength 1.550 µm are launched in the system at the input and add ports. The monochromatic optical field is given by

$$E_{in}(t) = E_0 \exp\left[-\alpha L + j\phi_0(t)\right] \tag{5.1}$$

Here $L = 2\pi R$ is the propagation distance (waveguide length), α the attenuation coefficient, and ϕ_0 the phase constant. When light propagates within the nonlinear material (medium), by considering the Kerr nonlinear effect within the ring devices, the refractive index (n) of light within the medium is given by

$$n = n_0 + n_2 I = n_0 + \frac{n_2}{A_{eff}} P \tag{5.2}$$

Here n_0 and n_2 are the linear and nonlinear refractive indices, respectively. I and P are the optical intensity and optical power, respectively. The effective mode core area of the device is given by A_{eff}.

The resonance output is formed, thus, the normalized output of the light field is the ratio between the output and input fields $[E_{out}(t)$ and $E_{in}(t)]$ in each roundtrip, which is given by the following equations [1, 4]:

$$\left|\frac{E_{out}(t)}{E_{in}(t)}\right|^2 = (1-\gamma) \cdot \left[1 - \frac{\left(1-(1-\gamma)x^2\right)\kappa}{\left(1-x\sqrt{1-\gamma}\sqrt{1-\kappa}\right)^2 + 4x\sqrt{1-\gamma}\sqrt{1-\kappa}\sin^2\left(\frac{\phi}{2}\right)}\right] \tag{5.3}$$

The optical output of ring resonator add/drop filter for the through and drop ports can be given by Eqs. 5.4 and 5.5, respectively [4].

$$\left|\frac{E_{t1}}{E_{in}}\right|^2 = \frac{(1-\kappa_1) - 2\sqrt{1-\kappa_1} \cdot \sqrt{1-\kappa_2} \cdot e^{\frac{-\alpha}{2}L} \cos(k_n L) + (1-\kappa_2)e^{-\alpha L}}{1 + (1-\kappa_1)(1-\kappa_2)e^{-\alpha L} - 2\sqrt{1-\kappa_1} \cdot \sqrt{1-\kappa_2}e^{\frac{-\alpha}{2}L} \cos(k_n L)} \tag{5.4}$$

$$\left|\frac{E_{t2}}{E_{in}}\right|^2 = \frac{\kappa_1 \kappa_2 \cdot e^{\frac{-\alpha}{2}L}}{1 + (1-\kappa_1)(1-\kappa_2)e^{-\alpha L} - 2\sqrt{1-\kappa_1} \cdot \sqrt{1-\kappa_2}e^{\frac{-\alpha}{2}L} \cos(k_n L)} \tag{5.5}$$

Here E_{t1} and E_{t2} represent the optical fields of the through and drop ports, respectively. $x = \exp(-\alpha L/2)$ is the roundtrip loss coefficient, $k_n = 2\pi/\lambda$ the wave propagation number in vacuum, n_{eff} the effective refractive index, $\varphi = kn_{eff}L$ the phase constant, γ the fractional coupler-intensity loss, κ the coupling coefficient, and β

the complex coefficient. The signals of both rings (of radii R_1 and R_2) are observed at the point Ref. 1 (E_{R1}) and Sen. 1 (E_{S1}), respectively, as shown in Fig. 5.1, and the mathematical forms of those signals are also analyzed, which can be expressed as

$$\left|\frac{E_{S1}}{E_{in}}\right|^2 = \left[\frac{-(1-\gamma_S)\kappa_S}{1-Z_2(1-\gamma_S)(1-\kappa_S)}\right] \cdot \left[\frac{j \cdot Z_3\sqrt{(1-\gamma_C)\kappa_C}\left(1+Z_3^2\beta_1\sqrt{(1-\gamma_C)(1-\kappa_C)}\right)}{1-Z_3^4\beta_1\beta_2(1-\gamma_C)(1-\kappa_C)}\right]^2$$

(5.6)

$$\left|\frac{E_{R1}}{E_{in}}\right|^2 = \left[\frac{-(1-\gamma_R)\kappa_R}{1-Z_1(1-\gamma_R)(1-\kappa_R)}\right] \cdot \left[\frac{j \cdot Z_3\sqrt{(1-\gamma_C)\kappa_C}\left(1+Z_3^2\beta_2\sqrt{(1-\gamma_C)(1-\kappa_C)}\right)}{1-Z_3^4\beta_1\beta_2(1-\gamma_C)(1-\kappa_C)}\right]^2$$

(5.7)

Here E_{S1} and E_{R1} are the sensing and reference signals, respectively; γ_S and γ_R are the fractional coupler-intensity losses in sensing and reference units, respectively; κ_S and κ_R are the coupling coefficients in sensing and reference units, respectively; $Z_1 = \exp\left(\dfrac{-\alpha}{8}\dfrac{L_1}{2} - jk_n\dfrac{L_1}{2}\right)$, $Z_2 = \exp\left(\dfrac{-\alpha}{8}\dfrac{L_2}{2} - jk_n\dfrac{L_2}{2}\right)$ and $Z_3 = \exp\left(\dfrac{-\alpha}{8}\dfrac{L_3}{4} - jk_n\dfrac{L_3}{4}\right)$ are loss coefficients, and β is the complex coefficient, which can be described as

$$\beta_1 = \left[\frac{\sqrt{(1-\gamma_3)(1-\kappa_3)}+(1-\gamma_3)\cdot e^{\frac{-\alpha L_1}{4} - jk_n L_1}}{1-\sqrt{(1-\gamma_3)(1-\kappa_3)}\cdot e^{\frac{-\alpha L_1}{4} - jk_n L_1}}\right]$$

(5.8)

$$\beta_2 = \left[\frac{\sqrt{(1-\gamma_0)(1-\kappa_0)}+(1-\gamma_0)\cdot e^{\frac{-\alpha L_2}{4} - jk_n L_2}}{1-\sqrt{(1-\gamma_0)(1-\kappa_0)}\cdot e^{\frac{-\alpha L_2}{4} - jk_n L_2}}\right]$$

(5.9)

The power output P at all ports is expressed by

$$P = |E|^2$$

(5.10)

To compare the reference and sensing signals, we set $\gamma_0 = \gamma_3$ and $\kappa_0 = \kappa_3$, so that $\beta_1 = \beta_2$ and then set $\gamma_S = \gamma_R$, so that $E_{S1} = E_{R1}$ when $L_1 = L_2$. E_{S1} varied while L_2 is changed by mean of varying R_2 with respect to E_{R1}, where R_1 remains constant. By using the finite difference time domain method (FDTD), the system is analyzed by using a computer program called "Opti-wave," whereas all parameters are simulated based on the practical parameters. The simulation steps are 40,000 iterations, and the peak spectrum at point Ref. 1 and Sen. 1 are set as reference and sensing signals, respectively, as shown in Figs. 5.2 and 5.3. In the figures, the change in the optical path length between the sensing and reference signals is compared and the change induced by the external parameters is measured.

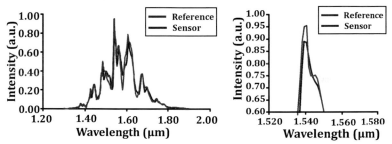

Figure 5.2 Graphs showing relationship between the intensity and wavelength of sensing (E_{S1}) and reference (E_{R1}) signals when the ring radii R_1 (*blue line*) and R_2 (*red line*) are 1.550 µm each.

This measurement is done by calculating the shift in wavelength ($\Delta\lambda$), i.e., the change in the optical path length, which is given by

$$\Delta\lambda = \lambda_2 - \lambda_1 \tag{5.11}$$

Here λ_1 and λ_2 are the peak wavelengths of Ref. 1 and Sen. 1, respectively. The relationships between the intensity and the wavelength shift are plotted in Fig. 5.3.

5.3 Nanoscale Sensing Operation

In simulation, the sensing ring radius R_2 is varied from 1.550 µm to 1.558 µm, due to which the optical path length changes and an interferometer system is formed [9, 10]. Both signals, i.e., sensing and reference, are observed, compared, and measured [5, 6]. We had assumed that the load cell or other sensing parameters can exert stress and strain on the second ring with radius R_2, but they actually appear in the sensing device due to the elastic modulus of the device material, which, in turn, is caused by the difference in peak spectrum of both signals and is described by Eq. 5.12.

$$Y_0 = \frac{F/A}{\Delta L/L} = \frac{\text{Stress}}{\text{Strain}} \tag{5.12}$$

The relationship between the force and the change in the length of the sensing device is described by

$$F = \left(\frac{Y_0 A_0}{L_0}\right) \cdot \Delta L \tag{5.13}$$

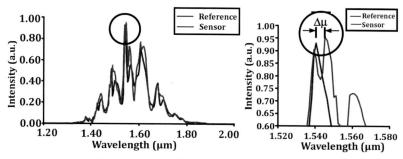

(a) R_1 (*blue line*) = 1.550 μm and R_2 (*red line*) = 1.552 μm

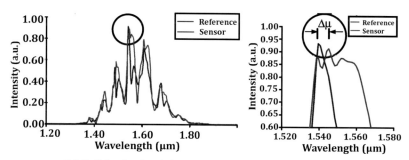

(b) R_1 (*blue line*) = 1.550 μm and R_2 (*red line*) = 1.554 μm

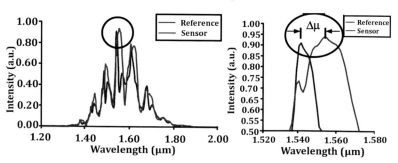

(c) R_1 (*blue line*) = 1.550 μm and R_2 (*red line*) = 1.556 μm

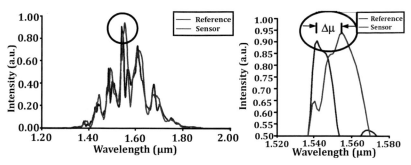

(d) R_1 (*blue line*) = 1.550 μm and R_2 (*red line*) = 1.558 μm

Figure 5.3 Graphs showing relationship between the intensity and wavelength of sensing (E_{S1}) and reference (E_{R1}) signals when the ring radii R_1 (*blue line*) = 1.550 μm and R_2 (*red line*) is equal to (a) 1.552 μm, (b) 1.554 μm, (c) 1.556 μm, and (d) 1.558 μm.

Here F is the applied force, Y_0 the Young modulus, A_0 the initial cross-section area, L_0 the initial length, and ΔL the change in length. The relationship between the force and the wavelength shift of InGaAsP/InP material [11–13] is plotted in Fig. 5.4.

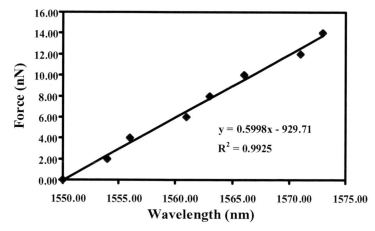

Figure 5.4 A graph of linear relationship between the force and the wavelength of the sensing signal.

By using Eq. 5.13 and the simulation results in Figs. 5.2 and 5.3, the relationship between the force and the wavelength shift of the reference and sensing signals is shown in Fig. 5.4. In this case, the nanoforce ranged between 0 and 16 nN and was caused by the coupling effect between the sensing device and the surrounding environment, such as molecules, DNA molecules, or atoms. We found that a sensing range in terms of wavelength shift ($\Delta\lambda$) is achieved within a resolution of 1 nm. Figure 5.3 shows the relationship between the varying ring radius (R_2) and the wavelength shift ($\Delta\lambda$) by comparing the sensing and reference signals in cases where self-calibration is formed. By using the least squares curve fitting, the linearity relationship between the applied force and the wavelength shift, with $R^2 = 0.9925$ is formed, which is a linearity for good sensing application. In addition, when force is applied on the sensing ring, where the change in resonance frequency, i.e., the free spectral range (FSR) is introduced, the ring radius R_2 varies from 1.550 μm to 1.558 μm. However, when the applied force is larger than 16.0 nN, a measurement limitation occurs due to the elastic limit of the sensing material and the plot of the relationship between the force and the wavelength shift no longer remains linear.

5.4   Conclusion

We proposed a nanoscale sensing system by using a PANDA ring resonator to take benefit of the accuracy of measurements in nanoscale range and a measurement resolution of 1 nm. Calibration is allowed by using the change in the wavelength between the sensing and reference signals, which exists within the system. Self-calibration of the measurement between the sensing and reference signals can be compared to each other without any additional optical part or unit. The other advantage of the proposed system is remote measurement of nanoscale force, which can be done due to the use of the integrated optic device.

References

1. Y. G. Boucher, and P. Féron, "Generalized transfer function: A simple model applied to active single-mode, microring resonators," *Opt. Commun.*, **282**, 3940–3947 (2009).

2. Y. Dumeige, C. Arnaud, and P. Féron, "Combining FDTD with coupled mode theories for bistability in micro-ring resonators," *Opt. Commun.*, **250**, 376–383 (2005).

3. T. T. Mai, F. Hsiao, C. Lee, W. Xiang, C. Chen, and W. K. Choi, "Optimization and comparison of photonic crystal resonators for silicon microcantilever sensors," *Sens. Actuators A*, **165**, 16–25 (2011).

4. K. Uomwech, K. Sarapat, and P. P. Yupapin, "Dynamic modutated Gaussian pulse propagation within the double PANDA ring resonator," *Microw. Opt. Technol. Lett.*, **52**(8), 1818–1821 (2010).

5. M. A. Haque, and M. T. A. Saif, "Application of MEMS force sensors for in situ mechanical characterization of nano-scale thin films in SEM and TEM," *Sens. Actuators A*, **A97–A98**, 239–245 (2002).

6. Z. Djinovic, M. Tomic, and A. Vujanic, "Nanometer scale measurement of wear rate and vibrations by fiber-optic white light interferometry," *Sens. Actuators A*, **A123–A124**, 92–98 (2005).

7. B. Piyatamrong, C. Teeka, R. Jomtarak, S. Mitatha, and P. P. Yupapin, "Multi photons trapping within optical vortices in an add/drop multiplexer," *Opt. Lett.* (2010) (in press).

8. T. Phatharaworamet, C. Teeka, R. Jomtarak, S. Mitatha, and P. P. Yupapin, "Random binary code generation using dark-bright soliton conversion control within a PANDA ring resonator," *IEEE J. Lightwave Technol.*, **28**(19), 2804–2809 (2010),

9. N. K. Berger, "Measurement of subpicosecond optical waveforms using a resonator-based phase modulator," *Opt. Commun.*, **283**, 1397–1405 (2010).

10. P. Hua, B. J. Luff, G. R. Quigley, J. S. Wilkinson, and K. Kawaguchi, "Integrated optical dual Mach-Zehnder interferometer sensor," *Sens. Actuators B*, **87**, 250–257 (2002).

11. M. Levinshtein, S. Rumyantsev, and M. Shur, "Handbook series on semiconductor parameters," Vol. 1, World Scientific, London, 147–168 (1996).

12. M. Levinshtein, S. Rumyantsev, and M. Shur, "Handbook series on semiconductor parameters," Vol. 2, World Scientific, London, pp. 153–179 (1999).

13. T. P. Pearsall (ed), "GaInAsP alloy semiconductors," John Wiley and Sons, New York (1982).

Chapter 6

Manipulation of Solar Energy Conversion

6.1 Introduction

Solar energy has become a practical alternative source of energy to fossil fuels. We should have efficient ways to convert photons into electricity, fuel, and heat. The need for better conversion technologies is a driving force behind many recent developments in biology [1, 2], materials [3], and, specially, nanoscience [4]. The sun has enormous untapped potential to supply for our growing energy needs. The barrier to the greater use of solar energy is its high cost compared to that of fossil fuels. However, this disparity in cost is decreasing with the rising prices of fossil fuels and the rising costs of mitigating their impact on the environment and climate. The high cost of solar energy is directly related to the low efficiency of the conversion methods, the modest energy density of solar radiation, and the costly materials currently required for the conversion. The development of materials and methods to improve solar energy conversion is primarily a scientific challenge: breakthroughs in the fundamental understanding ought to enable marked progress. There is plenty of room for improvement, since photovoltaic conversion efficiencies for inexpensive organic and dye-sensitized solar cells are currently about 10% or less, the conversion efficiency of photosynthesis is less than 1%, and the best solar thermal efficiency is 30%, and the theoretical limits suggest that we can do much better [5–7].

Solar conversion is a young science. Its major growth began in the 1970s, spurred by the oil crisis that highlighted the pervasive

Nanophotonics: Devices, Circuits, and Systems
Preecha P. Yupapin, Keerayoot Srinuanjan, and Surachart Kamoldilok
Copyright © 2013 Pan Stanford Publishing Pte. Ltd.
ISBN 978-981-4364-36-2 (Hardcover), 978-981-4364-37-9 (eBook)
www.panstanford.com

importance of energy to our personal, social, economic, and political lives [8, 9]. In contrast, fossil fuel science has developed for over more than 250 years, stimulated by the industrial revolution and the promise of abundant fossil fuels. The science of thermodynamics, for example, is intimately intertwined with the development of the steam engine. The Carnot cycle, the mechanical equivalent of heat, and entropy all played starring roles in the development of thermodynamics and the technology of heat engines. Solar energy science faces an equally rich future, with nanoscience [10] enabling the discovery of the guiding principles of photonic energy conversion and their use in the development of cost-competitive technologies.

Linear electronic devices and systems have been used in human life for years, and their various applications have been seen in the world. Till date, the progress in new technology and application is quite slow, and it may be limited in the near future. Therefore, the trend of new devices and technology has become an interesting subject of research today. Improvement in nonlinear device technology is one of the targets of research that should be encouraged in this era of new technologies. Some interesting results of nonlinear devices that have been reported [11, 12] have shown that interesting results can be obtained by using a nonlinear device known as a "nanoring resonator system." The use of this device in practical applications has also been reported [13]. Such a device was fabricated by using the nonlinear material InGaAsP/InP, where the nonlinear refractive index of this material is one of the key parameters of production. To begin this concept, we introduce the "ring resonator," a device that is in circular form or planar waveguide. Yupapin *et al.* [14, 15] have shown promising applications of this device by scaling down the ring radius of this device to within micrometer or nanometer range. For instance, ultrafast switching can be achieved by using a remarkably simple arrangement wherein a switching time of attosecond or more is achieved [11]. The most interesting result is when a light pulse can be slowed, stopped, and stored within a nonlinear nano-waveguide and signal amplification can be done within the tiny device, which is a great success. Recently, Yupapin and Suwancharoen have reported [16] interesting results of light pulse propagating within a nonlinear microring device, where the transfer function of the output at resonant condition is derived and used. They found that the broad spectrum of a light pulse can be transformed to discrete pulses. In this chapter, we propose an optical scheme that can be used to generate a continuous spectrum, which can be stored within the tiny device. First, the white light spectrum can be formed by using the soliton pulse within the novel system. Second, the selected pulse can be amplified and stored within the nano-waveguide. This system can be used to describe the concept of white-light generation and regeneration, which allows the collection and conversion of solar energy by using the designed nano-waveguide.

6.2 Soliton in a Nano-Waveguide System

An optical soliton is recognized as a powerful laser pulse, whose optical bandwidth can be enlarged when it is propagating within the nonlinear microring resonator. Moreover, the superposition of self-phase modulation soliton pulses can keep the large output power. Initially, optimum energy is coupled into the waveguide by a large effective core-area device, i.e., ring resonator, as shown in Fig. 6.1. Then, a smaller one is connected to form the stopping behavior. The filtering characteristic of the optical signal is presented within a ring resonator, where the suitable parameters are controlled to obtain the required output energy. To maintain the soliton pulse propagating within the ring resonator, a suitable coupling power is required in the device, The interference signal is a minor effect compared to the loss associated to the direct passing through. A soliton pulse is introduced into the multistage nanoring resonators as shown in Fig. 6.1, the input optical field (E_{in}) of the bright soliton pulse is given by Eq. 6.1 [17].

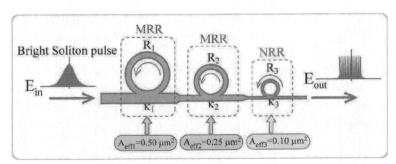

Figure 6.1   A schematic diagram of a continuous spectrum–generation system using a bright soliton. (R_s: ring radii, κ_s: coupling coefficient, MRR: microring resonator, NRR: nanoring resonator)

$$E_{in}(t) = A \operatorname{sech}\left[\frac{T}{T_0}\right]\exp\left[\left(\frac{z}{2L_D}\right) - i\omega_0 t\right] \quad (6.1)$$

Here A and z are the optical field amplitude and propagation distance, respectively. $T = t - \beta_1 Z$, where T is a soliton pulse propagation time in a frame moving at the group velocity, and $L_D = T_0^2/|\beta_2|$ is the dispersion length of the soliton pulse. β_1 and β_2 are the coefficients of the linear and second-order terms of the Taylor expansion of the propagation constant. T_0 in the equation is the initial soliton pulse width, where t is the soliton phase-shift time and x_0 the frequency shift of the soliton. This solution describes a pulse that keeps its temporal width invariance as it propagates, and is, thus, called a temporal soliton. When a soliton peak intensity ($|\beta_2/\Gamma T_0^2|$) is given, then T_0 is known. For the soliton pulse in the microring device, a balance should be achieved between the dispersion length (L_D) and the nonlinear length [$L_{NL} = (1/\Gamma\phi N_L)$], where $\Gamma = n_2 k_0$ is the length scale over which dispersive or nonlinear effects makes the beam becomes wider or narrower. For a soliton pulse, there is a balance between dispersion and nonlinear lengths, hence, $L_D = L_{NL}$.

When light propagates within the nonlinear material (medium), the refractive index (n) of light within the medium is given by n_0 and n_2, which are the linear and nonlinear refractive indexes, respectively. I and P are the optical intensity and optical power, respectively. The effective mode core area of the device is given by A_{eff}. For the microring and nanoring resonators, the effective mode core areas range from 0.50 μm² to 0.1 μm² [13], and a fast light pulse can be slowed down experimentally in this area after it enters the nanoring. When a soliton pulse is launched into and propagated within a microring resonator, which consists of a series microring resonators as shown in Fig. 6.1, the resonant output is formed when the optical wavelength matches with the optical path length. Therefore, the normalized output of the light field is the ratio between the output and input fields [$E_{out}(t)$ and $E_{in}(t)$] in each roundtrip, which can be expressed as follows [18]:

$$\left|\frac{E_{out}(t)}{E_{in}(t)}\right|^2 = (1-\gamma)\left[1 - \frac{(1-(1-\gamma)x^2)\kappa}{(1-x\sqrt{1-\gamma}\sqrt{1-\kappa})^2 + 4x\sqrt{1-\gamma}\sqrt{1-\kappa}\sin^2\left(\frac{\varphi}{2}\right)}\right]$$

$$(6.2)$$

The close form of Eq. 6.2 indicates that a ring resonator in this particular case is very similar to a Fabry–Perot cavity, which has an input and output mirror with a field reflectivity $1 - \kappa$ and a fully reflecting mirror. κ is the coupling coefficient, and $x = \exp(-\alpha L/2)$ represents a roundtrip loss coefficient, $\varphi_0 = kLn_0$ and $\phi_{NL} = kLn_2|E_{in}|^2$ are the linear and nonlinear phase shifts, $k = 2\pi/\lambda$ is the wave propagation number in a vacuum, where L and α are the waveguide length and linear absorption coefficient, respectively.

In this chapter, the iterative method is introduced to obtain the results as shown in Eq. 6.2, similarly, when the output field is connected and input into the other ring resonators.

6.3 White Light Generation and Amplification

In operation, a large bandwidth signal can be generated within the microring device by using a soliton pulse as input into the nonlinear microring resonator. This means that a white light spectra can be generated after the soliton pulse is input into the nonlinear microring resonator. A schematic diagram of the proposed system is shown in Fig. 6.1. A soliton pulse of 50 ns pulse width and 0.65 W peak power is input into the system. Suitable ring parameters are used, for instance, ring radii R_1 = 10.0 μm, R_2 = 7.0 μm, and R_3 = 5.0 μm. In order to make the system associate with the practical device [19, 20], the selected parameters of the system are fixed to k_0 = 1.55 μm, n_0 = 3.34 (InGaAsP/InP), A_{eff} = 0.50 μm², 0.25 μm² and 0.10 μm² for the microring and nanoring resonators [13], respectively, a = 0.5 dBmm⁻¹, and γ = 0.1.

The coupling coefficient (κ) of the microring resonator ranges from 0.03 to 0.1. The nonlinear refractive index n_2 = 2.2 × 10⁻¹³ m²/W. In this case, the wave-guided loss used is 0.5 dBmm⁻¹. The input soliton pulse is chopped (sliced) into smaller signals spreading over the spectrum (i.e., white light) as shown in Fig. 6.2, which shows that a large-bandwidth signal generates within the first ring (R_1) of the device. The biggest output amplification is obtained within the nano-waveguide (last ring, R_3). In Fig. 6.3, results of the relationship between output power/input power and wavelength are obtained when a soliton pulse is input into the ring resonator system as shown in Fig. 6.1, where the parameters used are R_1 = 10 μm, A_{eff1} = 0.50 μm², R_2 = 7 μm, A_{eff2} = 0.25 μm², R_3 = 5 μm, A_{eff3} = 0.10 μm²,

$\kappa_1 = 0.10$ and $\kappa_2 = \kappa_3 = 0.05$. A continuous spectra, which is 10 times larger than the input is obtained as output. The coupling coefficients are given as shown in the figures. We also found that the light pulse energy recovery (amplification) can be obtained by connecting the

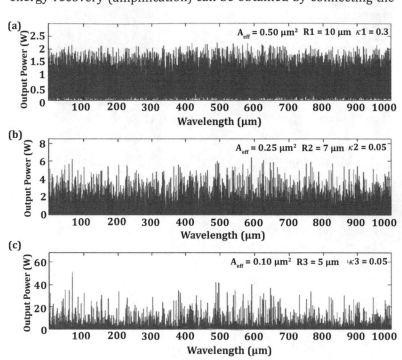

Figure 6.2 Results of the relationship between output power/input power and wavelength. Generation of large-bandwidth signals within rings (a) R_1, (b) R_2, and (c) R_3 (The biggest output amplification is obtained within this ring.).

nanoring device into the system. The coupling loss is included due to the different core effective areas between micro- and nanoring devices, which are given by the same value (0.1 dB). The key point of this proposal is that by using the nano-waveguide system, solar radiation, which is a low-level white light, can be collected and amplified. To improve the efficiency of solar energy collection is the challenge for today's technology. The continuous spectrum of a light pulse is chopped (sliced) to a discrete spectrum by the nonlinear behavior of light, which is known as chaotic property of light. The main parameter involved is the nonlinear refractive index of the device, i.e., the waveguide material and structure. The effective core area used is 0.10 μm², which is formed by a nano-waveguide. In principle, the power amplification within a nanoring device is obtained by the soliton behavior, known as self-phase and cross-phase modulations, when there is a balance between the dispersion and nonlinear length phase shift. This introduces the soliton pulse energy recovery, i.e., amplification, which generates the large white light output power. The solar energy can be stored within the nano-waveguide as white light spectra [12], which can be consumed even after a long time.

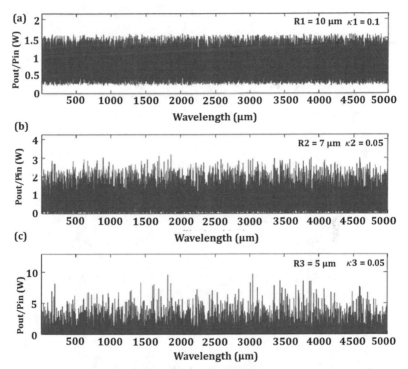

Figure 6.3 Results of the relationship between output power/input power and wavelength obtained within rings (a) R_1, (b) R_2, and (c) R_3 when a bright soliton pulse is input into the microring resonator system shown in Fig. 6.1.

6.4 Conclusion

We showed that a large bandwidth of optical signals with specific wavelength can be generated and amplified within a microring resonator system. The target for the future is to make a nanoscale device that convert and amplify solar energy and also generate and collect white light spectra, i.e., solar radiation. In Fig. 6.2, it was shown that a maximum power of 60 W can be obtained in ring R_3 by using the soliton pulse within a nano-waveguide. However, in Fig. 6.1, it was observed that coupling losses of different core effective areas of the waveguide occur, which affect the optical output in all cases.

References

1. D. Gust, T. Moore, and A. Moore, "Mimicking photosynthetic solar energy transduction," *Accounts Chem. Res.*, **34**(1), 40–48 (2001).

2. M. Grätzel, "Photoelectrochemical cells," *Nature*, **414**, 338–344 (2001).

3. Y. Hamakawa (ed), "Thin-film solar cells: Next generation photovoltaics and its applications," New York, Springer (2006).

4. A. J. Nozik, "Exciton multiplication and relaxation dynamics in quantum dots: Applications to ultrahigh-efficiency solar photon conversion," *Inorg. Chem.*, **44**(20), 6893–6899 (2005).

5. L. D. Hicks, and M. S. Dresselhaus, "Effect of quantum-well structures on the thermoelectric figure of merit," *Phys. Rev. B*, **47**, 12727–12731 (1993).

6. D. Mills, "Advances in solar thermal electricity technology," *Sol. Energy*, **76**, 19–31 (2004).

7. A. Steinfeld, "Solar thermochemical production of hydrogen — A review," *Sol. Energy*, **78**, 603–615 (2005).

8. M. A. Green, "Third-generation photovoltaics: Advanced solar energy conversion," Vol. 12, Springer Series in Photonics, New York, 160 (2006).

9. A. Heller, "Hydrogen-evolving solar cells," *Science*, **223**(4641), 1141–1148 (1984).

10. M. R. Wasielewski, "Energy, charge, and spin transport in molecules and self-assembled nanostructures inspired by photosynthesis," *J. Org. Chem.*, **71**(14), 5051–5066 (2006).

11. P. P. Yupapin, N. Pornsuwanchroen, and S. Chaiyasoonthorn, "Attosecond pulse generation using nonlinear microring resonators," *Microw. Opt. Technol. Lett.*, **50**, 3108–3111 (2008).

12. P. P. Yupapin, and N. Pornsuwancharoen, "Proposed arrangement of nonlinear microring resonator system for stopping and storing light pulse," *IEEE Photon Technol. Lett.*, **21**(6), 404–406 (2009).

13. Y. Su, F. Liu, and Q. Li, "System performance of show-light buffering and storage in silicon nano-waveguide," *Proc. SPIE*, **6783**, 68732 (2007).

14. P. P. Yupapin, and N. Pornsuwancharoen, "Guided wave optics and photonics: Microring resonator design for telephone network security," Nova Science Publishers, New York (2008).

15. P. P. Yupapin, N. Pornsuwancharoen, and J. Ali, "Optical soliton in microring resonator: Unexpected results and applications," Nova Science Publishers, New York (2009).

16. P. P. Yupapin, and W. Suwancharoen, "Chaotic signal generation and cancellation using a microring resonator incorporating an optical add/drop multiplexer," *Opt. Commun.*, **280**, 343–350 (2007).

17. G. P. Agarwal, "Nonlinear fiber optics," 4th ed., Academic Press, New York (2007).

18. N. Pornsuwancharoen, S. Chaiyasoonthorn, and P. P. Yupapin, "Fast and slow lights generation using chaotic signals in the nonlinear microring resonators for communication security," *Opt. Eng.*, **48**(1), 015002-1-5 (2008).

19. C. Fietz, and G. Shvets, "Nonlinear polarization conversion using microring resonators," *Opt. Lett.*, **32,** 1683–1685 (2007).

20. Y. Kokubun, Y. Hatakeyama, M. Ogata, S. Suzuki, and N. Zaizen, "Fabrication technologies for vertically coupled microring resonator with multilevel crossing busline and ultracompact-ring radius," *IEEE J. Sel. Top. Quantum Electron*, **11**, 4–10 (2005).

Chapter 7

Drug Delivery Manipulation

7.1   Introduction

Optical tweezers have been widely investigated in many research areas, such as biology, physics, engineering, and nanotechnology, especially, when there is a combination between nanomedicine and nanotechnology, which makes the research interesting. The use of nanotechnology in medicine for mesoscopic particle controlling and releasing has been reported by some scientists [1, 2]. Lee *et al.* [3] demonstrated the deformability test of a single red blood cell (RBC) by using a plastic optical trapping device in a microfluidic chip. In this test, they used the optical trapping techniques that were invented by Ashkin *et al.* [4].

Recently, promising techniques of nanoscopic volume trapping and transportation within an add/drop multiplexer have been reported in theory [5] and experiment [6]. The transporter in these techniques is known as an optical tweezer. The technique for generating optical tweezers has become a powerful tool for manipulating nanometer-sized particles. To date, many useful static tweezers have been well recognized and realized. Besides this, the use of dynamic tweezers has also been realized in practical works [7–9]. Schulz *et al.* [10] have shown that trapped atoms can be transferred between two optical potentials. Optical tweezers use forces exerted by intensity gradients in strongly focused beams of light to trap and move nanoscopic volumes of matter. During the movement, another combination of forces is induced between

Nanophotonics: Devices, Circuits, and Systems
Preecha P. Yupapin, Keerayoot Srinuanjan, and Surachart Kamoldilok
Copyright © 2013 Pan Stanford Publishing Pte. Ltd.
ISBN 978-981-4364-36-2 (Hardcover), 978-981-4364-37-9 (eBook)
www.panstanford.com

photons due to their interactions caused by photon scattering effects. The field intensity can also be adjusted and tuned to the desired gradient field and scattering force so that the suitable trapping force can be formed. Thus, by configuring the appropriated force for the transmitter/receiver part, nanoscopic particles can be transported over a long distance. In this chapter, the methods for generating the dynamic optical tweezers/vortices using a dark soliton, a bright soliton and a Gaussian pulse propagating within an add/drop optical multiplexer with two nanoring resonators (PANDA ring resonator) will be discussed. The dynamic behaviors of a soliton and Gaussian pulses are well described. [11]. By using the proposed system, the transceiver can be integrated to form a single device. The transceiver can be used to transport nanoscopic volumes of matter, especially, drugs.

7.2 Microscopic Volume Trapping Force Generation

Trapping forces are exerted by the intensity gradients in the strongly focused beams of light, which can be used to trap and move microscopic volumes of matter, in which the optical forces are customarily defined by the relationship [12].

$$F = \frac{Qn_{\mathrm{m}}P}{c} \tag{7.1}$$

Here Q is a dimensionless efficiency, n_{m} the index of refraction of the suspending medium, c the speed of light, and P the incident laser power for the specimen. Q represents the fraction of power utilized to exert force. For an incident plane wave on a perfectly absorbing particle, Q is equal to 1. To achieve stable trapping, the radiation pressure must create a stable, three-dimensional equilibrium. Because biological specimens are usually contained in aqueous medium, the dependence of F on n_{m} can rarely be exploited to achieve higher trapping forces. Increasing the laser power is possible, but only over a limited range due to the possibility of optical damage. Q itself is, therefore, the main determinant of the trapping force. It depends upon the numerical aperture (NA), the laser wavelength, the light polarization state, the laser mode structure, the relative index of refraction, and the geometry of the particle.

Furthermore, in the Rayleigh regime, trapping forces decompose naturally into two components. Since, in this limit, the electromagnetic field is uniform across the dielectric, particles can be treated as induced point dipoles. The scattering force is given by the following equation [12]:

$$F_{scatt} = n_m \frac{\langle S \rangle \sigma}{c}$$ (7.2)

where

$$\sigma = \frac{8}{3} \pi (kr)^4 r^2 \left(\frac{m^2 - 1}{m^2 + 2} \right)^2$$ (7.3)

Here σ is the scattering cross-section of a Rayleigh sphere with radius r. $\langle S \rangle$ is the time averaged Poynting vector, n the index of refraction of the particle, $m = n/n_m$ the relative index, and $k = 2\pi n_m/\lambda$ is the wave number of the light. The scattering force is proportional to the energy flux and points along the direction of propagation of the incident light. The gradient field (F_{grad}) is the Lorentz force acting on the dipole induced by the light field. It is given by the following equation [12]:

$$F_{grad} = \frac{\alpha}{2} \nabla \langle E^2 \rangle$$ (7.4)

where

$$\alpha = n_m^2 r^3 \left(\frac{m^2 - 1}{m^2 + 2} \right)$$ (7.5)

The polarizability of the particle *I* represented by α. The gradient force is proportional and parallel to the gradient in energy density (for $m > 1$). The large gradient force is formed by the large depth of the laser beam, in which the stable trapping requires the gradient force to be in the $-\hat{z}$ direction, which is against the direction of the incident light (dark soliton valley) and is greater than the scattering force. By increasing the NA and decreasing the focal spot size, the gradient strength can be increased [12] within a tiny system, such as a nanoscale device (nanoring resonator).

We are looking for a system that can generate dynamic tweezers (optical vortices) in which a microscopic volume can be trapped and transmitted via the communication link. First, a stationary and strong pulse is required that can propagate within the dielectric material (waveguide) for a given period of time. The gradient field is an important property required in this case. A dark soliton satisfies this condition and is, therefore, recommended for the pulse. Second, we are looking for a device that can propagate optical tweezers and form the long distance link, through which the gradient field (force) can be transmitted and received by using the same device. Here, the add/drop multiplexer is in the form of a PANDA ring resonator, which is well known and is being introduced for this proposal (Figs. 7.1 and 7.2). For multifunction operations, such as control, tune, and amplify, the additional pulses are bright soliton and Gaussian pulses introduced into the system. The input optical field (E_{in}) and the add port optical field (E_{add}) of the dark soliton, bright soliton and Gaussian pulses are given by [13], respectively.

$$E_{in}(t) = A_0 \tanh\left[\frac{T}{T_0}\right]\exp\left[\left(\frac{z}{2L_D}\right) - i\omega_0 t\right]$$ (7.6a)

$$E_{control}(t) = A \operatorname{sech}\left[\frac{T}{T_0}\right]\exp\left[\left(\frac{z}{2L_D}\right) - i\omega_0 t\right]$$ (7.6b)

$$E_{control}(t) = E_0 \exp\left[\left(\frac{z}{2L_D}\right) - i\omega_0 t\right]$$ (7.6c)

Here A and z are the optical field amplitude and propagation distance, respectively. $T = t - \beta_1 z$, where T is the soliton pulse propagation time in a frame moving at group velocity and $L_D = T_0^2/|\beta_2|$ is the dispersion length of the soliton pulse. β_1 and β_2 are the coefficients of the linear and second-order terms of the Taylor expansion of the propagation constant. T_0 is the soliton pulse propagation time at initial input (or soliton pulse width), where t is the soliton phase-shift time and ω_0 is the frequency shift of the soliton. This solution describes a pulse that keeps its temporal width invariance as it propagates and is, thus, called a temporal soliton. When a soliton of peak intensity $(|\beta_2/\Gamma T_0^2|)$ is given, then T_0 is known. For the soliton pulse in the microring device, a balance should be achieved between the dispersion length (L_D) and the nonlinear length $(L_{NL} = 1/\Gamma\phi_{NL})$, where $\Gamma = n_2 k_0$ is the length scale over which dispersive or nonlinear effects make the beam become wider or narrower. For a soliton pulse, there is a balance between dispersion and nonlinear lengths. Hence, $L_D = L_{NL}$. For a Gaussian pulse in Eq. 7.6c, E_0 is the amplitude of the optical field.

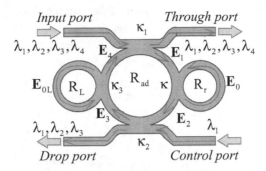

Figure 7.1   A schematic diagram of the proposed PANDA ring resonator.

When light propagates within a nonlinear medium, the refractive index (n) of light within the medium is given by

$$n = n_0 + n_2 I = n_0 + \frac{n_2}{A_{eff}} P \qquad (7.7)$$

Here n_0 and n_2 are the linear and nonlinear refractive indexes, respectively. I and P are the optical intensity and the power, respectively. The effective mode core area of the device is given by A_{eff}. For the add/drop optical filter design, the effective mode core areas range from 0.10 µm² to 0.50 µm², in which the parameters were obtained by using the related practical material parameters (InGaAsP/InP) [14]. When a dark soliton pulse is input and propagated within an add/drop optical filter as shown in Fig. 7.1, the resonant output is formed. Thus, the normalized output of the light field is defined as the ratio between the output and input fields [$E_{out}(t)$ and $E_{in}(t)$] in each roundtrip. This is given as [15].

$$\left| \frac{E_{out}(t)}{E_{in}(t)} \right|^2 = (1-\gamma) \left[1 - \frac{(1-(1-\gamma)x^2)\kappa}{(1-x\sqrt{1-\gamma}\sqrt{1-\kappa})^2 + 4x\sqrt{1-\gamma}\sqrt{1-\kappa} \sin^2\left(\frac{\phi}{2}\right)} \right]$$

$$(7.8)$$

The close form of Eq. 7.8 indicates that a ring resonator in this particular case is very similar to a Fabry–Perot cavity. The input and output mirror with a field reflectivity are $(1 - \kappa)$, and a fully reflecting mirror. κ is the coupling coefficient, and $x = \exp(-\alpha L/2)$ represents a roundtrip loss coefficient. $\phi_0 = kLn_0$ and $\phi_{NL} = kLn_2|E_{in}|^2$ are the linear and nonlinear phase shifts. $k = 2\pi/\lambda$ is the wave propagation number in a vacuum. L and α are the waveguide length and linear absorption coefficient, respectively. In this work, the iterative method is introduced to obtain resonance results in the same manner when the output field is connected and is input into other ring resonators.

In order to retrieve the required signals, we propose to use the add/drop device with the appropriate parameters. This is given in the following details. The optical circuits of the ring resonator add/drop filters for the through port and the drop port can be given by Eqs. 7.9 and 7.10, respectively [16].

$$\left|\frac{E_t}{E_{in}}\right|^2 = \frac{\left[\begin{array}{l}(1-\kappa_1)+(1-\kappa_2)e^{-\alpha L} \\ -2\sqrt{1-\kappa_1}\cdot\sqrt{1-\kappa_2}e^{-\frac{\alpha}{2}L}\cos(k_nL)\end{array}\right]}{\left[\begin{array}{l}1+(1-\kappa_1)(1-\kappa_2)e^{-\alpha L} \\ -2\sqrt{1-\kappa_1}\cdot\sqrt{1-\kappa_2}e^{-\frac{\alpha}{2}L}\cos(k_nL)\end{array}\right]} \quad (7.9)$$

$$\left|\frac{E_d}{E_{in}}\right|^2 = \frac{\kappa_1\kappa_2 e^{-\frac{\alpha}{2}L}}{1+(1-\kappa_1)(1-\kappa_2)e^{-\alpha L}-2\sqrt{1-\kappa_1}\cdot\sqrt{1-\kappa_2}e^{-\frac{\alpha}{2}L}\cos(k_nL)} \quad (7.10)$$

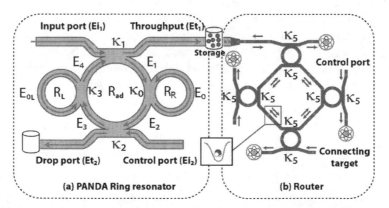

Figure 7.2 A schematic diagram of a drug delivery system using a transceiver and wavelength router, in which (a) is the PANDA ring resonator and (b) is the wavelength router.

Here E_t and E_d represent the optical fields of the through and drop ports, respectively. $\beta = kn_{eff}$ is the propagation constant,

n_{eff} the effective refractive index of the waveguide, and $L = 2\pi R$ the circumference of the ring, with R as the radius of the ring. The filtering signal can be managed by using specific parameters of the add/drop device, and the required signals can be retrieved via the drop port output. κ_1 and κ_2 are the coupling coefficients of the add/drop filters, $k_n = 2\pi/\lambda$ the wave propagation number in vacuum, and $\alpha = 0.5$ dBmm^{-1} the waveguide (ring resonator) loss. The fractional coupler intensity loss is $\gamma = 0.1$. In case of the add/drop device, the nonlinear refractive index does not affect the system, and is, therefore, neglected.

From Eq. 7.9, the output field E_{t1} at the through port is given by

$$E_{t1} = AE_{i1} - BE_{i2}e^{-\frac{\alpha L}{2} - jk_n\frac{L}{2}} - \left[\frac{CE_{i1}\left(e^{-\frac{\alpha L}{2} - jk_n\frac{L}{2}}\right)^2 + DE_{i2}\left(e^{-\frac{\alpha L}{2} - jk_n\frac{L}{2}}\right)^3}{1 - E\left(e^{-\frac{\alpha L}{2} - jk_n\frac{L}{2}}\right)^2}\right]$$

(7.11)

where, $A = \sqrt{(1-\gamma_1)(1-\gamma_2)}$, $B = \sqrt{(1-\gamma_1)(1-\gamma_2)\kappa_1(1-\kappa_2)}E_{0L}$, $C = \kappa_1(1-\gamma_1)\sqrt{(1-\gamma_2)\kappa_2}E_0E_{0L}$, $D = (1-\gamma_1)(1-\gamma_2)\sqrt{\kappa_1(1-\kappa_1)\kappa_2(1-\kappa_2)}E_0E_{0L}^2$, and $E = \sqrt{(1-\gamma_1)(1-\gamma_2)(1-\kappa_1)(1-\kappa_2)}E_0E_{0L}$.

The electric fields E_0 and E_{0L} are the field circulated within the nanoring at the right and left side of the add/drop optical filter.

The power output (P_{t1}) at the through port is written as

$$P_{t1} = |E_{t1}|^2$$

(7.12)

The output field (P_{t1}) at drop port is expressed as

$$E_{t2} = \sqrt{(1-\gamma_2)(1-\kappa_2)}E_{i2} - \left[\frac{\sqrt{(1-\gamma_1)(1-\gamma_2)\kappa_1\kappa_2}E_0E_{i1}e^{-\frac{\alpha L}{2} - jk_n\frac{L}{2}} + XE_0E_{0L}E_{i2}\left(e^{-\frac{\alpha L}{2} - jk_n\frac{L}{2}}\right)^2}{1 - YE_0E_{0L}\left(e^{-\frac{\alpha L}{2} - jk_n\frac{L}{2}}\right)^2}\right]$$

(7.13)

where $\quad X = (1 - \gamma_2)\sqrt{(1 - \gamma_1)(1 - \kappa_1)\kappa_2(1 - \kappa_2)},$

$$Y = \sqrt{(1 - \gamma_1)(1 - \gamma_2)(1 - \kappa_1)(1 - \kappa_2)}.$$

The power output (P_{t2}) at drop port is

$$P_{t2} = \left|E_{t2}\right|^2 \tag{7.14}$$

7.3 Microscopic Volume Transportation and Drug Delivery

Optical tweezers can be generated and trapped within the device and system as shown in Figs. 7.1 and 7.2. They can be used to transport

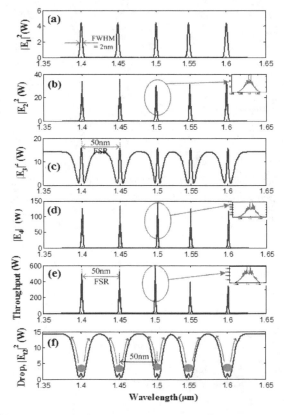

Figure 7.3 Results of dynamic tweezers with nanoscopic volumes (Fig. 7.2f), in which the generated wavelengths are 1.4 μm, 1.45 μm, 1.5 μm, 1.55 μm, and 1.6 μm.

and store the trapped volumes within the PANDA ring resonator and wavelength router, which can be used to transport nanoscopic volumes (in drug delivery) via the waveguide [6]. The manipulation of trapped nanoscopic volumes within the optical tweezers is shown in Fig. 7.3f. In this case study, the coupling coefficients are given as $\kappa_0 = 0.1$, $\kappa_1 = 0.35$, $\kappa_2 = 0.1$, and $\kappa_3 = 0.2$, respectively. The ring radii are $R_{add} = 1$ μm and 15 μm, $R_R = 100$ nm and 6 μm, $R_L = 100$ nm and 6 μm for which a device was reported by some authors [17]. A_{eff} are 0.50 μm², 0.25 μm², and 0.25 μm². In this case, dynamic tweezers (gradient fields) can be in the form of bright soliton, Gaussian pulses, and dark soliton, which can be used to trap the required nanoscopic volume.

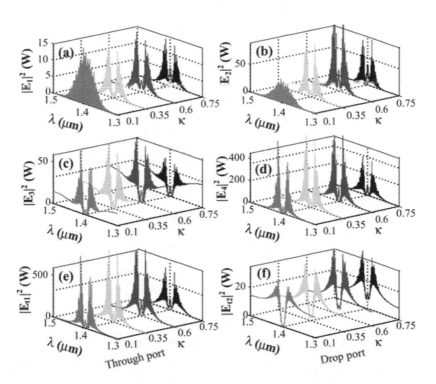

Figure 7.4 Simulation results of the tunable and amplified tweezers by varying the coupling coefficients.

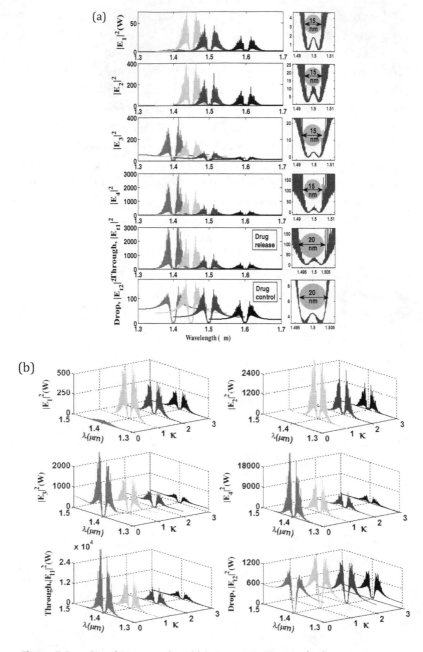

Figure 7.5 Simulation results of (a) the controlling and releasing drug into drop and throughput ports, respectively and (b) the tunable and amplified tweezers by varying the coupling coefficients.

Figure 7.4 shows the five different center wavelengths of tweezers that are generated when the dynamical movements are (a) $|E_1|^2$, (b) $|E_2|^2$, (c) $|E_3|^2$, (d) $|E_4|^2$, (e) through port signals, and (f) drop port signals. In this case, all nanoscopic volumes are received by the drop port. An important aspect of the result is that tweezers can be obtained by tuning (controlling) the add (control) port input signal, and the required number of nanoscopic volumes (atom/photon/molecule) can be obtained and seen at the drop/through ports, which otherwise, would propagate within the PANDA ring before collapsing/decaying into the waveguide. More results of the optical tweezers generated within the PANDA ring are shown in Fig. 7.5. In this case, the bright soliton is used as the control port signal. The output optical tweezers of the through and drop ports with different coupling constants are shown in Figs. 7.5e and f, respectively. The coupling coefficients are (1) 0.1, (2) 0.35, (3) 0.6, and (4) 0.75. The trapped nanoscopic volumes can be transported into the wavelength router via the through port, while the retrieved nanoscopic volumes can be received via the drop port (connecting target), which can perform the drug delivery applications. The advantage of the proposed system is that the transmitter and receiver can be fabricated on-chip and can be, alternatively, operated by a single device in liquid core, in which the use of light in waveguide comprises a liquid core and a liquid cladding [18]. The nanoscopic effect on a particle in liquid medium has been analyzed for narrow systems, which was given by references [19, 20]. The system is still effective in moving particles that are substantially smaller than the energizing wavelength, particularly, against the action of Brownian motion [21]. However, the magnitude of force can be varied and arranged by using tunable tweezer amplitudes as shown in Fig. 7.5, where tweezer widths of 20 nm can be obtained.

7.4 Conclusion

We showed that nanoscopic volumes can be trapped and transported into the optical waveguide with the help of optical tweezers. By using a PANDA ring resonator and wavelength router, the drugs can be transported over a long distance. By utilizing the reasonable dark soliton input power, dynamic tweezers can be controlled and stored within the system. Tweezers with free spectrum range

(FSR) and full width at half maximum (FWHM) of 50 nm and 2 nm, respectively, have been made. Tweezer amplification can also be obtained by using nanoring resonators and modulated signals via the control port as shown in Figs. 7.4 and 7.5. We also showed that a transceiver can be used for transporting nanoscopic volumes over a long distance by using the proposed system, in which drugs can be delivered via the wavelength router to the required (connecting) targets. However, problems of large nanoscopic volume and neutral matter may cause problems, and searching a new guide pipe and medium [22], such as a nanotube or a specific gas, will be the topic of investigation.

References

1. G. Huang, J. Gao, Z. Hua, J. V. St. John, B. C. Ponder, and D. Moro, "Controlled drug release from hydrogel nanoparticle networks," *J. Control. Release*, **94**, 303–311 (2004).

2. S. Acharya, and S. K. Sahoo, "PLGA nanoparticles containing various anticancer agents and tumour delivery by EPR effect," *Adv. Drug Deliv. Rev.*, **63**(3), 170–183 (2010).

3. W. G. Lee, K. Park, H. Bang, *et al.*, "Single red blood cell deformity test using optical trapping in plastic microfluid chip," *Proceedings of the 31 Annual International IEEE EMBS Special Topic on Conference Microtechnologies in Medicine and Biology Kahuku, Oahu, Hawaii, 12–15 May, 2005*, pp. 389–390.

4. A. Ashkin, J. M. Dziedzic, and T. Yamane, "Observation of a single-beam gradient force optical trap for dielectric," *Opt. Lett.*, **11**, 288–290 (1986).

5. B. Piyatamrong, K. Kulsirirat, S. Mitatha, and P. P. Yupapin, "Dynamic potential well generation and control using double resonators incorporating in an add/drop filter," *Mod. Phys. Lett. B.*, **24**, 3071–3082 (2010).

6. H. Cai and A. Poon, "Optical manipulation and transport of micro-particle on silicon nitride microring-resonator-based add-drop devices," *Opt. Lett.*, **35**, 2855–2857 (2010).

7. A. Ashkin, J. M. Dziedzic, J. E. Bjorkholm, and S. Chu, "Observation of a single beam gradient force optical trap for dielectric particles," *Opt. Lett.*, **11**, 288–290 (1986).

8. K. Egashira, A. Terasaki, and T. Kondow, "Photon trap spectroscopy applied to molecules adsorbed on a solid surface: Probing with a standing wave versus a propagating wave," *Appl. Opt.*, **80**, 5113–5115 (1998).

9. A. V. Kachynski, A. N. Kuzmin, H. E. Pudavar, D. S. Kaputa, A. N. Cartwright, and P. N. Prasad, "Measurement of optical trapping forces by use of the two-photon-excited fluorescence of microspheres," *Opt. Lett.*, **28**, 2288–2290 (2003).

10. M. Schulz, H. Crepaz, F. Schmidt-Kaler, J. Eschner, and R. Blatt, "Transfer of trapped atoms between two optical tweezer potentials," *J. Mod. Opt.*, **54**, 1619–1626 (2007).

11. M. Tasakorn C. Teeka, R. Jomtarak, and P. P. Yupapin, "Multitweezers generation control within a nanoring resonator system," *Opt. Eng.*, **49**, 075002-1-7 (2010).

12. K. Svoboda, and S. M. Block, "Biological applications of optical forces," *Annu. Rev. Biophys. Biomol. Struct.*, **32**, 247–282 (1994).

13. S. Mithata, N. Pornsuwancharoen, and P. P. Yupapin, "A simultaneous short wave and millimeter wave generation using a soliton pulse within a nano-waveguide," *IEEE Photon. Technol. Lett.*, **21**, 932 934 (2009).

14. Y. Kokubun, Y. Hatakeyama, M. Ogata, S. Suzuki, and N. Zaizen, "Fabrication technologies for vertically coupled microring resonator with multilevel crossing busline and ultracompact-ring radius," *IEEE J. Sel. Top. Quantum Electron.*, **11**, 4–10 (2005).

15. P. P. Yupapin, and W. Suwancharoen, "Chaotic signal generation and cancellation using a micro ring resonator incorporating an optical add/drop multiplexer," *Opt. Commun.*, **280**(2), 343–350 (2007).

16. P. P. Yupapin, P. Saeung, and C. Li, "Characteristics of complementary ring-resonator add/drop filters modeling by using graphical approach," *Opt. Commun.*, **272**, 81–86 (2007).

17. J. Zhu, S. K. Ozdemir, Y. F. Xiao, *et al.*, "On-chip single nanoparticle detection and sizing by mode splitting in an ultrahigh-Q microresonator," *Nat. Photonics*, **4**, 46–49 (2010).

18. D. B. Wolfe, R. S. Conroy, P. Garsteck, *et al.*, "Dynamic control of liquid-core liquid-cladding optical waveguides," *Proc. Natl. Acad. Sci.*, **101**(34), 2434–12438 (2004).

19. L. Tongcang, S. Kheifets, D. M. Edellin, and M. G. Raizen, "Measurement of the instantaneous velocity of a Brownian particle," *Science*, **328**(25), 1673–1675 (2010).

20. H. S. Chung, and R. Hogg, "The effect of Brownian motion on particle size analysis by sedimentation," *Powder Tech.*, **41**, 211–216 (1985).

21. P. Zemanek, A. Jonas, L. Sramek, and M. Liska, "Optical trapping of Rayleigh particles using a Gaussian standing wave," *Opt. Commun.*, **151**, 273–285 (1998).

22. A. Ashkin, "Optical trapping and manipulation of neutral particles using lasers," *Proc. Natl. Acad. Sci.*, **94**, 4853–4860 (1997).

Chapter 8

Molecular Buffer

8.1 Introduction

Optical buffer is recognized as an essential component in a wavelength router, in which packets of data can be stored for resolving packet contention problem and also to delay the outgoing packets [1–2]. In practice, optical router patents have been proposed and recorded [3–5], which can be useful for various applications. Recently, promising techniques of microscopic volume trapping and transportation within an add/drop multiplexer have been reported in both theory [6] and experiment [7], respectively. The transporter in these techniques is known as an optical tweezer. The technique for generating optical tweezers has become a powerful tool for manipulating of micrometer-sized particles. Useful static tweezers have now been well recognized and realized. The use of dynamic tweezers has also been realized in practical works [8–10]. Schulz *et al.* [11] have shown that trapped atoms between can be transported between two optical potentials. Optical tweezers use forces exerted by intensity gradients in strongly focused beams of light to trap and move nanoscopic volumes of matter. During the movement, another combination of forces is induced between photons due to their interactions caused by photon scattering effects. The field intensity can also be adjusted and tuned to the desired gradient field and scattering force so that the suitable trapping force can be formed. Thus, by configuring the appropriated force for the transmitter/

Nanophotonics: Devices, Circuits, and Systems
Preecha P. Yupapin, Keerayoot Srinuanjan, and Surachart Kamoldilok
Copyright © 2013 Pan Stanford Publishing Pte. Ltd.
ISBN 978-981-4364-36-2 (Hardcover), 978-981-4364-37-9 (eBook)
www.panstanford.com

receiver part, nanoscopic particles can be transported over a long distance.

In this chapter, the methods for generating the dynamic optical tweezers/vortices using a dark soliton, a bright soliton and a Gaussian pulse propagating within an add/drop optical multiplexer with two nanoring resonators (PANDA ring resonator) will be discussed. The dynamic behaviors of a soliton and Gaussian pulses are well described. [12]. By using the proposed system, the transceiver can be integrated to form a single device. The transceiver can be used to transport nanoscopic volumes of matter [13], especially, molecules in liquid core waveguide [14, 15], drugs, and DNA, where a buffer is needed before the substance reaches its destination.

8.2 Theoretical Background

Trapping forces are exerted by the intensity gradients in the strongly focused beams of light, which can be used to trap and move microscopic volumes of matter, in which the optical forces are customarily defined by the relationship [16].

$$F = \frac{Q n_m P}{c} \tag{8.1}$$

Here Q is a dimensionless efficiency, n_m the index of refraction of the suspending medium, c the speed of light, and P the incident laser power for the specimen. Q represents the fraction of power utilized to exert force. For an incident plane wave on a perfectly absorbing particle, Q is equal to 1. To achieve stable trapping, the radiation pressure must create a stable, three-dimensional equilibrium. Because biological specimens are usually contained in aqueous medium, the dependence of F on n_m can rarely be exploited to achieve higher trapping forces. Increasing the laser power is possible, but only over a limited range due to the possibility of optical damage. Q itself is, therefore, the main determinant of the trapping force. It depends upon the numerical aperture (NA), the laser wavelength, the light polarization state, the laser mode structure, the relative index of refraction, and the geometry of the particle.

Furthermore, in the Rayleigh regime, trapping forces decompose naturally into two components. Since, in this limit, the electromagnetic field is uniform across the dielectric, particles can be treated as induced point dipoles. The scattering force is given by the following equation [16]:

$$F_{scatt} = n_m \frac{\langle S \rangle \sigma}{c} \tag{8.2}$$

where

$$\sigma = \frac{8}{3} \pi (kr)^4 r^2 \left(\frac{m^2 - 1}{m^2 + 2} \right)^2 \tag{8.3}$$

Here σ is the scattering cross-section of a Rayleigh sphere with radius r. $\langle S \rangle$ is the time averaged Poynting vector, n the index of refraction of the particle, $m = n/n_m$ the relative index, and $k = 2\pi n_m/\lambda$ is the wave number of the light. The scattering force is proportional to the energy flux and points along the direction of propagation of the incident light. The gradient field (F_{grad}) is the Lorentz force acting on the dipole induced by the light field. It is given by the following equation [16]:

$$F_{grad} = \frac{\alpha}{2} \nabla \langle E^2 \rangle \tag{8.4}$$

where

$$\alpha = n_m^2 r^3 \left(\frac{m^2 - 1}{m^2 + 2} \right) \tag{8.5}$$

The polarizability of the particle is α. The gradient force is proportional and parallel to the gradient in energy density (for $m > 1$). The large gradient force is formed by the large depth of the laser beam, in which the stable trapping requires the gradient force to be in the $-\hat{z}$ direction, which is against the direction of the incident light (dark soliton valley) and is greater than the scattering force. By increasing the NA and decreasing the focal spot size, the gradient strength can be increased [16, 17], within a tiny system, such as a nanoscale device (nanoring resonator).

In our proposal, the trapping force is formed by using a dark soliton, in which a valley of the dark soliton is generated and controlled within the PANDA ring resonator by the control port signals. In this chapter, we will use the same theory of optical trapping and ring resonance, but the simulation results and applications will differ from the previous work [18].

From Fig. 8.1, the output field (E_{t1}) at the through port is given by the following equation [19]:

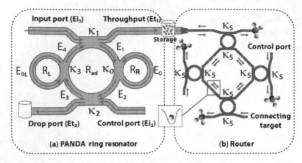

(a) PANDA ring resonator   (b) Router

Figure 8.1 A schematic diagram of a buffer system, in which (a) is the PANDA ring resonator and (b) is the wavelength router. R_{add} is the add/drop filter radius and R_R and R_L are the right and left ring resonator radii, respectively.

$$E_{t1} = AE_{i1} - BE_{i2}e^{-\frac{\alpha L}{2}-jk_n\frac{L}{2}} - \left[\frac{CE_{t1}\left(e^{-\frac{\alpha L}{2}-jk_n\frac{L}{2}}\right)^2 + DE_{i2}\left(e^{-\frac{\alpha L}{2}-jk_n\frac{L}{2}}\right)^3}{1-F\left(e^{-\frac{\alpha L}{2}-jk_n\frac{L}{2}}\right)^2}\right]$$

(8.6)

where, $A = \sqrt{(1-\gamma_1)(1-\gamma_2)}$, $B = \sqrt{(1-\gamma_1)(1-\gamma_2)\kappa_1(1-\kappa_2)}E_{0L}$, $C = \kappa_1(1-\gamma_1)\sqrt{(1-\gamma_2)\kappa_2}E_0E_{0L}$, $D = (1-\gamma_1)(1-\gamma_2)\sqrt{\kappa_1(1-\kappa_1)\kappa_2(1-\kappa_2)}E_0E_{0L}^2$, and $F = \sqrt{(1-\gamma_1)(1-\gamma_2)(1-\kappa_1)(1-\kappa_2)}E_0E_{0L}$.

Here E_t and E_d represent the optical fields of the through port and drop ports, respectively. $\beta = kn_{eff}$ is the propagation constant, n_{eff} the effective refractive index of the waveguide, and $L = 2\pi R$ the circumference of the ring, where R is the radius of the ring. κ_1 and κ_2 are the coupling coefficients of the add/drop filters, $k_n = 2\pi/\lambda$ the wave propagation number in vacuum, and $\alpha = 0.5$ dBmm^{-1} the waveguide (ring resonator) loss. The fractional coupler intensity loss is $\gamma = 0.1$. In the case of an add/drop device, the nonlinear refractive index does not affect the system, therefore, it is neglected. The electric fields E_0 and E_{0L} are the fields circulated within the nanoring at the right and left side of an add/drop optical filter.

The power output (P_{t1}) at through port is written as

$$P_{t1} = |E_{t1}|^2$$

(8.7)

The output field (E_{t2}) at the drop port is expressed by the following equation [18]:

$$E_{t2} = \sqrt{(1-\gamma_2)(1-\kappa_2)}E_{i2} - \left[\cfrac{\sqrt{(1-\gamma_1)(1-\gamma_2)\kappa_1\kappa_2}E_0E_{i1}e^{-\frac{\alpha L}{2}-jk_n\frac{L}{2}} + XE_0E_{0L}E_{i2}\left(e^{-\frac{\alpha L}{2}-jk_n\frac{L}{2}}\right)^2}{1 - YE_0E_{0L}\left(e^{-\frac{\alpha L}{2}-jk_n\frac{L}{2}}\right)^2} \right]$$

(8.8)

where $X = (1-\gamma_2)\sqrt{(1-\gamma_1)(1-\kappa_1)\kappa_2(1-\kappa_2)}$ and

$$Y = \sqrt{(1-\gamma_1)(1-\gamma_2)(1-\kappa_1)(1-\kappa_2)}$$

The power output (P_{t2}) at drop port is

$$P_{t2} = |E_{t2}|^2$$

(8.9)

Optical tweezers can be trapped, transported, and stored within a PANDA ring resonator and wavelength router, which can then be used to transport microscopic volumes of molecules or drugs via the waveguide [19].

8.3 Molecular Buffer

A molecular buffer is a device that can be used to store or delay atoms/molecules for a period of time, in which the intensity and velocity of light can also be controlled (Fig. 8.2) [20, 21]. This device can be used in medical applications. A molecular buffer is a device, which is operated in the same way as the buffer for gases [22]. The polarizability of a particle is calculated by using Eq. 8.5. In this case, we assume that the particle is polystyrene (n = 1.5894), the liquid medium is water (n = 1.33), and the optical power which is required to trap particles of a certain size/polarizability is 9.1 W, which is represented by the slope of the curve shown in Fig. 8.3a.

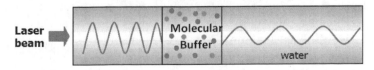

Figure 8.2   A schematic diagram of a molecular buffer working in a core waveguide.

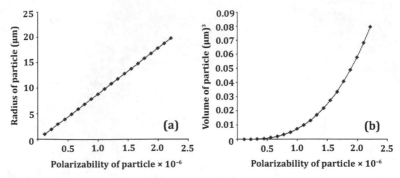

Figure 8.3 Graphs of optical powers which are required to trap particles of certain size and polarizability.

In simulation, the bright soliton with a center wavelength at 1.50 μm, peak power 2 W, and pulse 35 fs is input into the system via the input port. The coupling coefficients are given as $\kappa_0 = 0.5$, $\kappa_1 = 0.35$, $\kappa_2 = 0.1$, and $\kappa_3 = 0.35$, respectively. The ring radii are $R_{add} = 10$ μm and 30 μm, $R_R = 50$ nm and 100 nm, and $R_L = 50$ nm and 100 nm, respectively. A practically feasible device with a radius of 30 nm has been reported by some authors [18]. A_{eff} are 0.50 μm², 0.25 μm², and 0.25 μm². In this case, dynamic tweezers (gradient fields) can be in the form of bright solitons, Gaussian pulses, and dark solitons, which can be used to trap the required microscopic volume. Four different center wavelengths of tweezers have been generated, and their dynamical movements are shown in Fig. 8.4 for (a) $|E_1|^2$, (b) $|E_2|^2$, (c) $|E_3|^2$, (d) $|E_4|^2$, (e) through port signals, and (f) drop port signals. In this case, all microscopic volumes are received by the drop port. Fabrication parameters that can be easily controlled are not the coupling constants but the ring resonator radii.

An important aspect of the result is that tweezers can be obtained by tuning (controlling) the add (control) port input signal, and the required number of nanoscopic volumes (atom/photon/molecule) can be obtained and seen at the drop/through ports, which otherwise, would propagate within the PANDA ring before collapsing/decaying into the waveguide. More results of the optical tweezers generated within the PANDA ring are shown in Fig. 8.5. In this case, the bright soliton is used as the control port signal to obtain tunable results. The output optical tweezers of the through and drop ports with different coupling constants are shown in Fig. 8.5a, while results with different wavelengths are shown in Fig. 8.5b, which can be performed for selected targets. The trapped microscopic volumes (molecules) can transport into the wavelength router via the through port, while the retrieved microscopic volumes are received via the drop port (connecting target). The advantage of the proposed system is that the transmitter and receiver can be fabricated on-chip and, alternatively, can be operated by a single device.

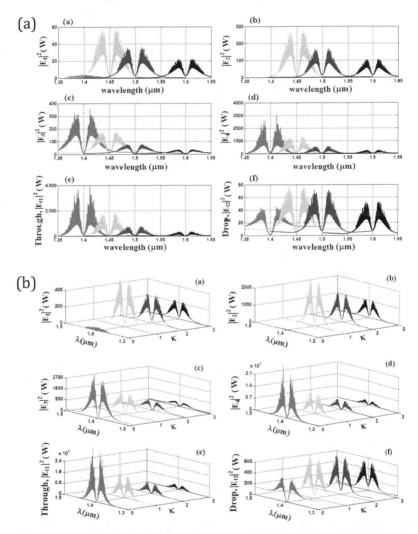

Figure 8.4   Results of dynamic tweezers within the buffer with different (a) wavelengths and (b) coupling constants, where $R_{add} = 10\ \mu m$, $R_R = R_L = 100$ nm.

The optical trapping force is of pico-Newton (pN) magnitude and depends upon the relative refractive index of the particle [23]. The particle radius is located in the cavity [24–26]. The radius decreases with a decrease in the refractive index of the host medium.

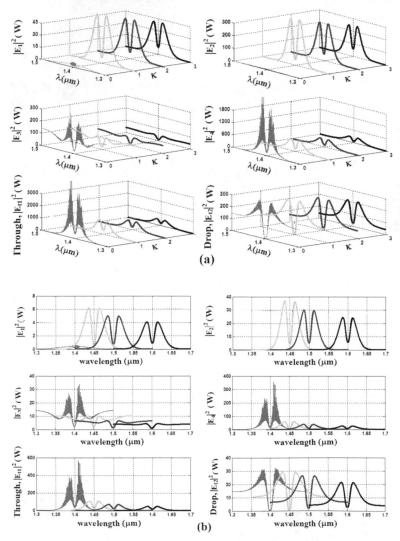

Figure 8.5 Results of dynamic tweezers within the buffer with different (a) coupling constants and (b) wavelengths, where $R_{add} = 30\,\mu m$, $R_R = R_L = 50$ nm.

8.4 Conclusion

In this chapter, we have proposed a new system that can be used to trap (delay) and transport molecules into an optical waveguide by using optical tweezers, which can be used for drug storage and as a drug-delivery system. By utilizing the reasonable dark soliton input power, dynamic tweezers can be controlled and stored (delayed) within the system before they reach the final destination. Tweezers can also be amplified by using nanoring resonators, in which signals can be modulated via the control port as shown in Figs. 8.4b and 8.5a. We also showed that a transceiver can be used to transport microscopic volumes over a long distance by using the proposed system, in which drugs can be delivered via the wavelength router to the required (connecting) targets. However, problems of large nanoscopic volume and neutral matter may cause problems, and searching a new guide pipe and medium [27], such as a nanotube or a specific gas, will be the topic of investigation.

References

1. M. Cheng, C. Wu, J. Hiltunen, Y. Wang, Q. Wang, and R. Myllylä, "A variable delay optical buffer based on nonlinear polarization rotation in semiconductor optical amplifier," *IEEE Photon. Technol. Lett.*, **21**, 1885–1887 (2009).

2. J. Liu, T. T. Lee, X. Jiang, and S. Horiguchi, "Blocking and delay analysis of single wavelength optical buffer with general packet-size distribution," *J. Lightwave Technol.*, **27**, 955–966 (2009).

3. C. P. Dragone, "Improved waveguide grating optical router suitable for CWDM," US patent EP2250523. March 9, 2010.

4. B. S. Ham, "Delayed optical router/switch," US patent 20100232792. September 16, 2010.

5. K. Oguchi, and S. Terada, "Optical network system, optical router, fault recovery method of optical network, and program," US patent JP2010063009, March 18, 2010.

6. B. Piyatamrong, K. Kulsirirat, S. T. Mitatha, and P. P. Yupapin, "Dynamic potential well generation and control using double resonators incorporating in an add/drop filter," *Mod. Phys. Lett. B*, **24**, 3071–3082 (2010).

7. H. Cai, and A. Poon, "Optical manipulation and transport of microparticle on silicon nitride microring resonator-based add-drop devices," *Opt. Lett.*, **35**, 2855–2857 (2010).

8. A. Ashkin, J. M. Dziedzic, J. E. Bjorkholm, and S. Chu, "Observation of a single-beam gradient force optical trap for dielectric particles," *Opt. Lett.*, **11**, 288–290 (1986).

9. K. Egashira, A. Terasaki, and T. Kondow, "Photon-trap spectroscopy applied to molecules adsorbed on a solid surface: Probing with a standing wave versus a propagating wave," *Appl. Opt.*, **80**, 5113–5115 (1998).

10. A. V. Kachynski, A. N. Kuzmin, H. E. Pudavar, D. S. Kaputa, A. N. Cartwright, and P. N. Prasad, "Measurement of optical trapping forces by the use of two-photon-excited fluorescence of microspheres," *Opt. Lett.*, **28**, 2288–2290 (2003).

11. M. Schulz, H. Crepaz, F. Schmidt-Kaler, J. Eschner, and R. Blatt, "Transfer of trapped atoms between two optical tweezer potentials," *J. Mod. Opt.*, **54**, 1619–1626 (2007).

12. M. Tasakorn, C. Teeka, R. Jomtarak, and P. P. Yupapin, "Multitweezers generation control within a nanoring resonator system," *Opt. Eng.*, **49**, 075002 (2010).

13 K. Neuman and S. M. Block, "Optical trapping," *Rev. Sci. Instrum.*, **75**, 2787–2809 (2004).

14. S. J. Parkin, G. Knöner, T. A. Nieminen, N. R. Heckenberg, and H. R. Dunlop, "Picoliter viscometry using optically rotated particles," *Phys. Rev. E*, **76**, 041507-1-5 (2007).

15. D. B. Wolfe, R. S. Conroy, P. Garsteck, *et al.*, "Dynamic control of liquid-core liquid-cladding optical waveguides," *Proc. Natl. Acad. Sci.*, **101**(34), 2434–12438 (2004).

16. K. Svoboda, and S. M. Block, "Biological applications of optical forces," *Annu. Rev. Biophys. Biomol. Struct.*, **23**, 247–282 (1994).

17. J. Zhu, S. K. Ozdemir, Y. F. Xiao, L. Li, L. He, and D. R. Chen, "On-chip single nanoparticle detection and sizing by mode splitting in an ultrahigh-Q microresonator," *Nat. Photonics*, **4**, 46–49 (2010).

18. N. Suwanpayak, M. A. Jalil, T. Teeka, J. Ali, and P. P. Yupapin, "Optical vortices generated by a PANDA ring resonator for drug trapping and delivery applications," *Biomed. Opt. Express*, **2**(1), 159–168 (2011).

19. B. Piyatamrong, K. Kulsirirat, W. Techitdheera, S. Mitatha, and P. P. Yupapin, "Dynamic potential well generation and control using double resonators incorporating in an add/drop filter," *Mod. Phys. Lett. B*, **24**(32), 3071–3082 (2010).

20. M. A. Rosenberry, J. P. Reyes, D. Tupa, and T. J. Gay, "Radiation trapping in rubidium optical pumping at low buffer-gas pressures," *Phys. Rev. A*, **75**, 023401-1-6 (2007).

21. M. C. Lignie, and J. P. Woerdman, "Light-induced drift of Na in molecular buffer gases," *J. Phy. B: At. Mol. Opt. Phys.*, **23**, 417–426 (1990).

22. P. S. Waggoner, J. S. Palmer, V. N. Antonov, and J. H. Weaver, "Metal nanostructure growth on molecular buffer layers of CO_2," *Surf. Sci.*, **596**, 12–20 (2005).

23. R. Kumar, C. Shakher, and D. S. Mehtac, "3D multiple optical trapping of Au nanoparticles and prokaryote *E. coli* using intra-cavity generated non-circular beam of inhomogeneous intensity," *Laser Phys.*, **20**(6), 1514–1524 (2010).

24. J. Hu, S. Lin, L. C. Kimerling, and K. Crozier, "Optical trapping of dielectric nanoparticles in resonant cavities," *Phys. Rev. A.*, 0538191-8 (2010).

25. M. Fischer and K. B. Sørensen, "Calibration of trapping force and response function of optical tweezers in viscoelastic," *J. Opt. A: Pure Appl. Opt.*, **79**, 239–250 (2007).

26. T. A. Nieminen, H. R. Dunlop, and N. R. Heckenberg, "Calculation and optical measurement of laser trapping forces on non-spherical particles," *J. Quant. Spectrosc. Radiat. Transfer*, **70**, 627–637 (2001).

27. A. Ashkin, "Optical trapping and manipulation of neutral particles using lasers," *Proc. Natl. Acad. Sci.*, **94**, 4853–4860 (1997).

Chapter 9

Nanobattery Manipulation

9.1   Introduction

Nanobatteries have been widely studied and investigated by researchers [1–4] for their small size, long life, and high efficiency. To date, many types of thin film batteries (nanobatteries) have been investigated and reported [5–11]. Ferreira *et al.* [5] have reported the use of cellulose paper as electrolyte, separation of electrodes, and physical support simultaneously of a rechargeable battery. The use of Vanadia xerogel nanocathodes has been reported in lithium microbatteries [6]. The rate of charging these batteries is faster than the rate lithium [7] can homogenize in an active particle by diffusion, in which the inhomogeneous distribution of lithium results in stresses that may cause the particle to fracture. The electronic structure of $LiFeSO_4F$ [8] have been shown to be an excellent material as a cathode for lithium-ion batteries. Hu *et al.* [9] have reported that the characteristics thin and flexible are secondary to Li-ion paper batteries. Li-ion batteries have contributed to the commercial success of portable electronics and are now in a position to influence higher-volume applications such as plug-in hybrid electric vehicles, which have been well described by some authors [10]. A thin film [11–14] can be used in many fuel fields, such as solar cells. A thin film solar cell [15, 16] can be used to trap light. Many researchers have reported characteristics and applications of thin films [17–23].

From the above discussion, it is clear that the expected device should be small in size, be easy to use, and have a long-life operation.

Nanophotonics: Devices, Circuits, and Systems
Preecha P. Yupapin, Keerayoot Srinuanjan, and Surachart Kamoldilok
Copyright © 2013 Pan Stanford Publishing Pte. Ltd.
ISBN 978-981-4364-36-2 (Hardcover), 978-981-4364-37-9 (eBook)
www.panstanford.com

An optical thin film battery is a good candidate for such a device [5]. However, the requirement of a nanobattery is that it should have a construction technique that is reliable. Recently, Suwanpayak *et al.* [24] reported the use of a PANDA ring resonator for drug trapping and delivery applications. In this device, the photons/atoms/ molecules can be trapped and moved within an optical router. In this chapter, we will present the use of a trapping tool for moving atoms/ molecules for their delivery and collection within a thin film grating, where finally, the wavelength-selected tweezers will be formed. An atom assembly within the desired thin film grating can be made by using a PANDA microring resonator. The design for a thin film grating and the simulation result for it have been discussed in this chapter.

9.2   Theoretical Background

By using the optical trapping tool generated from a PANDA ring resonator [24–26], in which the specified trapping tools (tweezers) can be desired and controlled, atoms/molecules can be moved and embedded to the specified grating layers.

To start this concept, the proposed system consists of an add/ drop filter and double nanoring resonators as shown in Fig. 9.1a. For the trapping tools to perform, dark and bright solitons are input into the add/drop optical filter system, while the input optical field (E_{i1}) and the control port optical field (E_{con}) of the dark-bright solitons pulses are given by

$$E_{i1}(t) = A \tanh\left[\frac{T}{T_0}\right] \exp\left[\left(\frac{z}{2L_D}\right) - i\phi(t)\right] \tag{9.1}$$

$$E_{con}(t) = A \operatorname{sech}\left[\frac{T}{T_0}\right] \exp\left[\left(\frac{z}{2L_D}\right) - i\phi(t)\right] \tag{9.2}$$

Here A and z are the optical field amplitude and propagation distance, respectively. $\phi(t) = \phi_0 + \phi_{NL} = \phi_0 + \dfrac{2\pi n_2 L}{A_{eff}\lambda}|E_0(t)|^2$ is the random phase term related to the temporal coherence function of the input light, ϕ_0 the linear phase shift, ϕ_{NL} the nonlinear phase shift, and n_2 the nonlinear refractive index of InGaAsP/InP waveguide. The effective mode core area of the device is given by $A_{eff} L = 2\pi R_{ad}$, where R_{ad} is the radius of device, λ the input wavelength light field, and $E_0(t)$ the circulated field within the nanoring coupled to the right and left add/drop optical filter system as shown in Fig. 9.1. $T = t - \beta_1 z$, where T is the soliton pulse propagation time in a frame moving at the group velocity and $L_D = T_0^2/|\beta_2|$ is the dispersion length of the soliton pulse. β_1 and β_2 are the coefficients of the linear and second-order terms of the Taylor expansion of the propagation constant. T_0 is the soliton pulse propagation time at initial input (or soliton pulse width), where t is the soliton phase-shift time and the angular frequency shift of the soliton is ω_0. This solution describes a pulse that keeps its temporal width invariance as it propagates and is, thus, called a temporal soliton. When a soliton peak intensity $(|\beta_2/\Gamma T_0^2|)$ is given, then T_0 is known. For a soliton pulse in a microring device, a balance should be achieved between the dispersion length (L_D) and the nonlinear length $(L_{NL} = 1\Gamma\phi_{NL})$, where $\Gamma = n_2 k_n$ is the length scale over which dispersive or nonlinear effects make the beam wider or narrower. For a soliton pulse, there is a balance between dispersion and nonlinear lengths, hence, $L_D = L_{NL}$.

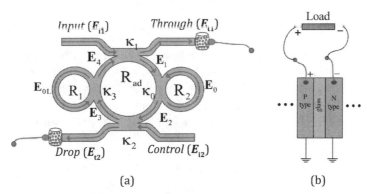

Figure 9.1   A schematic diagram of an atom bottom-up assembly, where (a) is a PANDA microring resonator pumped and controlled by light [27] and (b) is a bottom-up two-cell assembly of P- and N-type design for a nanobattery.

When light propagates within the nonlinear medium, the refractive index (n) of light within the medium is given by

$$n = n_0 + n_2 I = n_0 + \frac{n_2}{A_{eff}} P$$

Here n_0 and n_2 are the linear and nonlinear refractive indexes, respectively. I and P are the optical intensity and power, respectively.

The effective mode core area of the device is given by A_{eff}. For the add/drop optical filter design, the effective mode core areas range from 0.50 μm² to 0.10 μm² and the parameters were obtained by using the related practical material parameters InGaAsP/InP. When a dark soliton pulse is input and propagated within an add/drop optical filter, as shown in Fig. 9.1, the resonant output is formed. The resonator output fields, E_{t1} and E_1, consist of transmitted and circulated components within the add/drop optical multiplexing system, which can provide the driving force to photon/molecule/atom.

When the input light pulse passes through the first coupling device of the add/drop optical multiplexing system, the transmitted and circulated components can be written as

$$E_{t1} = \sqrt{1-\gamma_1}\left[\sqrt{1-\kappa_1}E_{i1} + j\sqrt{\kappa_1}E_4\right] \tag{9.3}$$

$$E_1 = \sqrt{1-\gamma_1}\left[\sqrt{1-\kappa_1}E_4 + j\sqrt{\kappa_1}E_{i1}\right] \tag{9.4}$$

$$E_2 = E_0 E_1 e^{-\frac{\alpha L}{2}\frac{L}{2} - jk_n \frac{L}{2}} \tag{9.5}$$

Here κ_1 is the intensity coupling coefficient, γ_1 the fractional coupler intensity loss, α the attenuation coefficient, $k_n = 2\pi/\lambda$ is the wave propagation number, λ the input wavelength light field, and $L = 2\pi R_{ad}$, where R_{ad} is the radius of the add/drop device.

For the second coupler of the add/drop optical multiplexing system,

$$E_{t2} = \sqrt{1-\gamma_2}\left[\sqrt{1-\kappa_2}E_{i2} + j\sqrt{\kappa_2}E_2\right] \tag{9.6}$$

$$E_3 = \sqrt{1-\gamma_2}\left[\sqrt{1-\kappa_2}E_2 + j\sqrt{\kappa_2}E_{i2}\right] \tag{9.7}$$

$$E_4 = E_{0L} E_3 e^{-\frac{\alpha L}{2}\frac{L}{2} - jk_n \frac{L}{2}} \tag{9.8}$$

Here κ_2 is the intensity coupling coefficient and γ_2 the fractional coupler intensity loss. The circulated light fields, E_0 and E_{0L}, are the light field circulated components of the nanorings with radii R_r and R_L, which are coupled into the right and left sides of the add/drop optical multiplexing system, respectively. The transmitted and circulated components of the light field in the right nanoring are given by

$$E_2 = \sqrt{1-\gamma}\left[\sqrt{1-\kappa_0}E_1 + j\sqrt{\kappa_0}E_{r2}\right] \tag{9.9}$$

$$E_{r1} = \sqrt{1-\gamma}\left[\sqrt{1-\kappa_0}E_{r2} + j\sqrt{\kappa_0}E_1\right] \tag{9.10}$$

$$E_{r2} = E_{r1}e^{-\frac{\alpha}{2}L_1 - jk_n L_1} \tag{9.11}$$

Here κ_0 is the intensity coupling coefficient, γ the fractional coupler intensity loss, α the attenuation coefficient, $k_n = 2\pi/\lambda$ the wave propagation number, λ the input wavelength light field, and $L_1 = 2\pi R_r$, R_r is the radius of the right nanoring.

From Eqs. 9.9–9.11, the circulated roundtrip light fields of the right nanoring with radii R_r are given by Eqs. 9.12 and 9.13, respectively.

$$E_{r1} = \frac{j\sqrt{1-\gamma}\sqrt{\kappa_0}E_1}{1-\sqrt{1-\gamma}\sqrt{1-\kappa_0}e^{-\frac{\alpha}{2}L_1 - jk_n L_1}} \tag{9.12}$$

$$E_{r2} = \frac{j\sqrt{1-\gamma}\sqrt{\kappa_0}E_1 e^{-\frac{\alpha}{2}L_1 - jk_n L_1}}{1-\sqrt{1-\gamma}\sqrt{1-\kappa_0}e^{\frac{\alpha}{2}L_1 \ jk_n L_1}} \tag{9.13}$$

Thus, the output circulated light field, E_0, for the right nanoring is given by

$$E_0 = E_1 \left\{ \frac{\sqrt{(1-\gamma)(1-\kappa_0)}-(1-\gamma)e^{-\frac{\alpha}{2}L_1 - jk_n L_1}}{1-\sqrt{(1-\gamma)(1-\kappa_0)}e^{-\frac{\alpha}{2}L_1 - jk_n L_1}} \right\} \tag{9.14}$$

Similarly, the output circulated light field, E_{0L}, for the left nanoring on the left side of the add/drop optical multiplexing system is given by

$$E_{0L} = E_3 \left\{ \frac{\sqrt{(1-\gamma_3)(1-\kappa_3)}-(1-\gamma_3)e^{-\frac{\alpha}{2}L_2 - jk_n L_2}}{1-\sqrt{(1-\gamma_3)(1-\kappa_3)}e^{-\frac{\alpha}{2}L_2 - jk_n L_2}} \right\} \tag{9.15}$$

Here κ_3 is the intensity coupling coefficient, γ_3 the fractional coupler intensity loss, α the attenuation coefficient, $k_n = 2\pi/\lambda$ the wave propagation number, λ the input wavelength light field, and $L_2 = 2\pi R_L$, where R_L is the radius of the left nanoring.

From Eqs. 9.3–9.15, the circulated light fields, E_1, E_3, and E_4 are defined by given $x_1 = (1-\gamma_1)^{1/2}$, $x_2 = (1-\gamma_2)^{1/2}$, $y_1 = (1-\kappa_1)^{1/2}$, and $y_2 = (1-\kappa_2)^{1/2}$.

$$E_1 = \frac{jx_1\sqrt{\kappa_1}E_{i1} + jx_1x_2y_1\sqrt{\kappa_2}E_{0L}E_{i2}e^{-\frac{\alpha L}{2}\frac{L}{2} - jk_n\frac{L}{2}}}{1 - x_1x_2y_1y_2E_0E_{0L}e^{-\frac{\alpha}{2}L - jk_nL}}$$ (9.16)

$$E_3 = x_2y_2E_0E_1e^{-\frac{\alpha L}{2}\frac{L}{2} - jk_n\frac{L}{2}} + jx_2\sqrt{\kappa_2}E_{i2}$$ (9.17)

$$E_4 = x_2y_2E_0E_{0L}E_1e^{-\frac{\alpha}{2}L - jk_nL} + jx_2\sqrt{\kappa_2}E_{0L}E_{i2}e^{-\frac{\alpha L}{2}\frac{L}{2} - jk_n\frac{L}{2}}$$ (9.18)

Thus, from Eqs. 9.3, 9.5, 9.16–9.18, the output optical field of the through port (E_{t1}) is expressed as

$$E_{t1} = x_1y_1E_{i1} + \left(\frac{jx_1x_2y_2\sqrt{\kappa_1}E_0E_{0L}E_1}{-x_1x_2\sqrt{\kappa_1\kappa_2}E_{0L}E_{i2}}\right)e^{-\frac{\alpha L}{2}\frac{L}{2} - jk_n\frac{L}{2}}$$ (9.19)

The power output of the through port (P_{t1}) is written as

$$P_{t1} = (E_{t1})\cdot(E_{t1})^* = |E_{t1}|^2$$ (9.20)

Similarly, from Eqs. 9.5, 9.6, 9.16–9.18, the output optical field of the drop port (E_{t2}) is given by

$$E_{t2} = x_2y_2E_{i2} + jx_2\sqrt{\kappa_2}E_0E_1e^{-\frac{\alpha L}{2}\frac{L}{2} - jk_n\frac{L}{2}}$$ (9.21)

The power output of the drop port (P_{t2}) is expressed as

$$P_{t2} = (E_{t2})\cdot(E_{t2})^* = |E_{t2}|^2$$ (9.22)

In Fig. 9.1b, a bottom-up two-cell assembly of P- and N-type based on a dielectric mirror (also known as a Bragg reflector) is shown. This assembly consists of identical alternating layers of high and low refractive indices and in which the number of cells can be increased by connecting to series or parallel schemes. The optical thicknesses are typically chosen to be a quarter of the wavelength, that is, $n_H l_H = n_L l_L = \lambda_0/4$ at some operating wavelength $\lambda_0 = 1550$ nm. The standard arrangement is to have an odd number of layers, with the high index layer being the first and the last layer.

After the first layer, we may view the structure as a repetition of N identical bilayers of low and high index. The elementary reflection coefficients alternate in sign (P- and N-types) as shown in Fig. 9.1b and are given by

$$\rho = \frac{n_{\rm H} - n_{\rm L}}{n_{\rm H} + n_{\rm L}}, \quad -\rho = \frac{n_{\rm L} - n_{\rm H}}{n_{\rm L} + n_{\rm H}} \tag{9.23}$$

The substrate $n_{\rm b}$ can be arbitrary, even the same as the incident medium $n_{\rm a}$. In that case, $\rho_2 = \rho_1$. The reflectivity properties of the structure can be understood by propagating the impedances from bilayer to bilayer. For example, in Fig. 9.1b, we have the quarter wavelength as

$$Z_2 = \frac{\eta_{\rm L}^2}{Z_3} = \frac{\eta_{\rm L}^2}{\eta_{\rm H}^2} Z_4 = \left(\frac{n_{\rm H}}{n_{\rm L}}\right)^2 Z_4 = \left(\frac{n_{\rm H}}{n_{\rm L}}\right)^4 Z_6 = \left(\frac{n_{\rm H}}{n_{\rm L}}\right)^6 Z_8 = \left(\frac{n_{\rm H}}{n_{\rm L}}\right)^8 \eta_{\rm b} \tag{9.24}$$

Therefore, after each bilayer, the impedance decreases by a factor of $(n_{\rm L}/n_{\rm H})^2$. After N bilayers, we will have

$$Z_2 = \left(\frac{n_{\rm H}}{n_{\rm L}}\right)^{2N} \eta_{\rm b} \tag{9.25}$$

Using $Z_1 = \eta_{\rm H}^2/Z_2$, the reflection response at λ_0 can be expressed as

$$R = \frac{Z_1 - \eta_{\rm a}}{Z_1 + \eta_{\rm a}} = \frac{1 - \left(\frac{n_{\rm H}}{n_{\rm L}}\right)^{2N} \frac{n_{\rm H}^2}{n_{\rm a} n_{\rm b}}}{1 + \left(\frac{n_{\rm H}}{n_{\rm L}}\right)^{2N} \frac{n_{\rm H}^2}{n_{\rm a} n_{\rm b}}} \tag{9.26}$$

Generally, it follows that for large N, R will tend to -1, that is, 100 % reflection.

9.3 Atom Trapping and Bottom-Up

By using the proposed design, atoms/molecules can be trapped and transported via the optical waveguide [24–26]. In this work, a thin film nanobattery, as shown in Fig. 9.1, can be constructed by using the atom bottom-up assembly. Simulation results of the atom trapping manipulation within the PANDA ring are shown in Fig. 9.2. In this case, the Gaussian modulated CW is input into the control port. The parameters of a PANDA microring are $R_{\rm ad} = 10$ μm, $R_1 = 4$ μm, and $R_2 = 4$ μm and practical evidence of this device have been reported by some authors [27]. In this case, the dynamic light pulses (tweezers) can be in the form of Gaussian pulses as shown in Fig. 9.2. The dynamic pulse generated at center wavelength 1.55 μm can be used to trap the required atom. The transported pulses are shown in Fig. 9.2, where (a) is the through port signal and (b) is the drop port

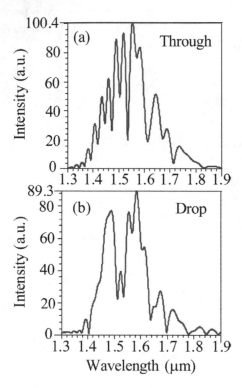

Figure 9.2   Simulation results of the optical trapping tool using a PANDA microring resonator for (a) through port signals and (b) drop port signals.

signal. In Fig. 9.3, the selected wavelength tweezers are generated and recorded within a thin film grating and the use of pulses for positive and negative trappings is shown. Fig. 9.3a,b show the positive pole (P-type) and negative pole (N-type), respectively. The trapped P-type and N-type atoms are required to embed within the specified thin film grating. By using the through port signal for trapping, as shown in Fig. 9.2a, the bottom-up two-cell assembly of P- and N-types of thin film nanobattery design can be constructed. The selected tweezers and atoms can be used to allocate the required thin film layers. In Fig. 9.4, we found that the transmitted light trapping tool (tweezer) at the wavelength 1.64 µm (Fig. 9.4a) and the reflected wavelength at 1.525 µm (Fig. 9.4b) are seen. The atom trapping and embedding can be used to assemble atoms within the thin film grating, in which trapping and assembly can also be done in case of many atoms (Fig. 9.5). In Fig. 9.6, the selected wavelength tweezers for trapping and confining the single-atom bottom-up assembly controlled by light are shown. Figure 9.6a shows simulation results at the selected wavelength of 1.5800 µm. In the proposed system, a single-atom is trapped and confined within a thin film nanobattery by selecting wavelengths of 1.5811 µm and 1.6524 µm for the P-type and N-type, respectively. The trapped P-type and N-type atoms can be embedded within the thin film grating layers by using the transporter and a thin film battery can be formed.

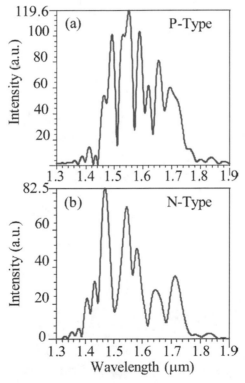

Figure 9.3 Light trapping manipulation results obtained at (a) positive pole (P-type) and (b) negative pole (N-type) by using a PANDA microring pump as shown in Fig. 9.2.

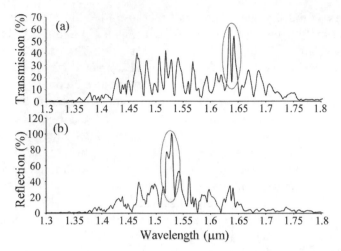

Figure 9.4 Transmission and reflection of bottom-up two-cell assembly of P- and N-type design at (a) 1.64 μm and (b) 1.525 μm, respectively.

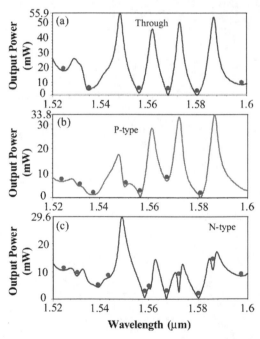

Figure 9.5 Dynamic results of atom bottom-up assembly controlled by light when output is at (a) through port, (b) P-type, and (c) N-type.

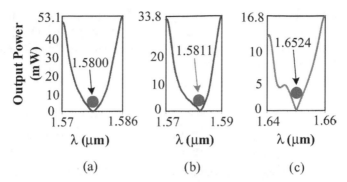

Figure 9.6 Simulation results of atom bottom-up assembly controlled by light for (a) through port pump signal at 1.5800 μm, (b) single atom trapping at 1.5811 μm for P-type, and (c) single atom trapping at 1.6524 μm for N-type.

9.4 Conclusion

In this chapter, we have proposed the use of a trapping tool for atom trapping, delivery, and embedding, which can be later trapped within a thin film grating. The selected atoms/molecules can be trapped and moved and then filtered and embedded within the required thin film grating layers, where finally, the nanobattery operation can be realized. The required atoms (P-type or N-type) can be trapped and confined within the required region (layers), where finally, the nanobattery construction, based on thin film grating, can be pumped and controlled by light by using a PANDA microring resonator.

References

1. F. Vullum, and D. Teeters, "Investigation of lithium battery nanoelectrode arrays and their component nanobatteries," *J. Power Sources*, **146**, 804–808 (2005).

2. A. Brazier, L. Dupont, L. Dantras-Laffont, N. Kuwata, J. Kawamura, and J. M. Tarascon, "First cross-section observation of an all solid-state lithium-ion "nanobattery" by transmission electron microscopy," *Chem. Mater.*, **20**, 2352–2359 (2008).

3. K. Zhao, M. Pharr, J. J. Vlassak, and Z. Suo, "Inelastic hosts as electrodes for high-capacity lithium-ion batteries," *J. Appl. Phys.*, **109**, 016110 (2011).

4. F. Vullum, D. Teeters, A. Nyten, and J. Thomas, "Characterization of lithium nanobatteries and lithium battery nanoelectrode arrays that benefit from nanostructure and molecular self-assembly," *Solid State Ionics*, **177**, 2833–2838 (2006).

5. I. Ferreira, B. Brás, N. Correia, P. Barquinha, E. Fortunato, and R. Martins, "Self-rechargeable paper thin-film batteries: Performance and applications," *J. Disp. Technol.*, **6**(8), 332–335 (2010).

6. C. Dewan and D. Teeters, "Vanadia xerogel nanocathodes used in lithium microbatteries," *J. Power sources*, **119–121**, 310–315 (2003).

7. K. Zhao, M. Pharr, J. J. Vlassak, and Z. Suo, "Fracture of electrodes in lithium-ion batteries caused by fast charging," *J. Appl. Phys.*, **108**, 073517 (2010).

8. M. Ramzan, S. Lebegue, and R. Ahuja, "Crystal and electronic structures of lithium fluorosulphate based materials for lithium-ion batteries," *Phys. Rev. B*, **82**, 125101 (2010).

9. L. Hu, H. Wu, F. L. Mantia, Y. Yang, and Y. Cui, "Thin, flexible secondary Li-ion paper batteries," *ACS Nano*, **4**(10), 5843–5848 (2010).

10. N. Recham, J. N. Chotard, L. Dupont, *et al.*, "A 3.6 V lithium-based fluorosulphate insertion positive electrode for lithium-ion batteries," *Nat. Mater.*, **9**, 68–74 (2010).

11. F. Wang, H. Y. Yu, J. Li, X. Sun, X. Wang, and H. Zheng, "Optical absorption enhancement in nanopore textured-silicon thin film for photovoltaic application," *Opt. Lett.*, **35**(1), 40–42 (2010).

12. Q. Hu, J. Wang, Y. Zhao, and D. Li, "A light-trapping structure based on Bi_2O_3 nano-islands with highly crystallized sputtered silicon for thin-film solar cells," *Opt. Express*, **19**(S1), A20–A27 (2011).

13. I. Diukman, L. Tzabari, N. Berkovitch, N. Tessler, and M. Orenstein, "Controlling absorption enhancement in organic photovoltaic cells by patterning Au nano disks within the active layer," *Opt. Express*, **19**(S1), A64–A71 (2011).

14. Y. A. Akimov, W. S. Koh, and K. Ostrikov, "Enhancement of optical absorption in thin-film solar cells through the excitation of higher-order nanoparticle plasmon modes," *Opt. Express*, **17**(12), 10195–10205 (2009).

15. V. E. Ferry, M. A. Verschuuren, H. B. T. Li, *et al.*, "Light trapping in ultrathin plasmonic solar cells," *Opt. Express*, **18**(S2), A237–A245 (2010).

16. R. Dewan, and D. Knipp, "Light trapping in thin-film silicon solar cells with integrated diffraction grating," *J. Appl. Phys.*, **106**, 074901 (2009).

17. S. H. Ahn, and L. J. Guo, "Dynamic nanoinscribing for continuous and seamless metal and polymer nanogratings," *Nano Lett.*, **9**(12), 4392–4397 (2009).

18. P. Genevet, J. P. Tetienne, E. Gatzogiannis, *et al.*, "Large enhancement of nonlinear optical phenomena by plasmonic nanocavity gratings," *Nano Lett.*, **10**, 4880–4883 (2010).

19. J. N. Munday, and H. A. Atwater, "Large integrated absorption enhancement in plasmonic solar cells by combining metallic gratings and antireflection coatings," *Nano Lett.*, **11**(6), 2195–2201 (2011).

20. Y. Zhu, and J. M. Tour, "Graphene nanoribbon thin films using layer-by-layer assembly," *Nano Lett.*, **10**, 4356–4362 (2010).

21. J. H. Cho, and D. H. Gracias, "Self-assembly of lithographically patterned nanoparticles," *Nano Lett.*, **9**(12), 4049–4052 (2009).

22. G. A. Kalinchenko, and A. M. Lerer, "Wideband all-dielectric diffraction grating on chirped mirror," *J. Lightwave Technol.*, **28**(18), 2743–2749 (2010).

23. V. N. Konopsky and E. V. Alieva, "Long-range propagation of plasmon polaritons in a thin metal film on a one-dimensional photonic crystal surface," *Phys. Rev. Lett.*, **97**, 253904 (2006).

24. N. Suwanpayak, M. A. Jalil, C. Teeka, J. Ali, and P. P. Yupapin, "Optical vortices generated by a PANDA ring resonator for drug trapping and delivery applications," *Biomed. Opt. Express*, **2**(1), 159–168 (2011).

25. P. Youplao, T. Phattaraworamet, S. Mitatha, C. Teeka, and P. P. Yupapin, "Novel optical trapping tool generation and storage controlled by light," *J. Nonlinear Opt. Phys. Mater.*, **19**(2), 371–378 (2010).

26. M. Tasakorn, C. Teeka, R. Jomtarak, and P. P. Yupapin, "Multitweezers generation control within a nanoring resonator system," *Opt. Eng.*, **49**(7), 075002 (2010).

27. J. Zhu, S. K. Ozdemir, Y. F. Xiao, *et al.*, "On-chip single nanoparticle detection and sizing by mode splitting in an ultrahigh-Q microresonator," *Nat. Photonics*, **4**, 46–49 (2010).

Chapter 10

Blood Cleaner On-Chip Design

10.1 Introduction

The kidneys are important excretory organs of human body and they help in maintaining concentrations of various ions and other important substances and in removing body wastes through urine [1]. A human kidney has approximately 300,000 to 1,000,000 nephrons [2], depending upon the relationship between weight at birth and the number and size of renal glomeruli in humans [3]. Each nephron is composed of two parts, the glomerulus and the tubules. The glomerulus is built up of capillary vessels, and blood plasma is filtered from blood through the porous wall of the vessels [1].

Chronic kidney disease (CKD) is a worldwide health problem, which is increasing dramatically in several countries [4, 5]. This is the one of the key areas where health care planning is needed even in developed countries, where many clinical organizations have been developed for the manufacture of devices and development of methods for hemodialysis [6–8]. CKD and acute renal failure occur when patients suffers from gradual and usually permanent loss of kidney function over time and lose their ability to filter and remove wastes and extra fluids from their bodies.

Hemodialysis is the process of removing waste products from the blood. This removal is normally done by the kidneys, but when they fail to do so [9], a dialysis machine (artificial kidney) performs the function. An artificial kidney was developed as a human kidney and is implanted when the human kidney [10, 11] fails to work.

Nanophotonics: Devices, Circuits, and Systems
Preecha P. Yupapin, Keerayoot Srinuanjan, and Surachart Kamoldilok
Copyright © 2013 Pan Stanford Publishing Pte. Ltd.
ISBN 978-981-4364-36-2 (Hardcover), 978-981-4364-37-9 (eBook)
www.panstanford.com

However, the artificial kidney dialysis has several disadvantages, such as line wastage from blood during dialysis, which takes about 2–4 hours time to clear. For this, the patient has to visit the hospital two to three times in a week. This led researchers and inventors to develop implantable artificial kidney [15], which was convenient for the patients.

Optical trapping was first discovered by Ashkin [16]. It emerged as a powerful tool and had broad applications in biology, physics, engineering, and medicine [17]. The ability of optical trapping and manipulating viruses, living cells, and bacteria by laser radiation pressure without destroying the organelles [18] being treated has been demonstrated in some experiments [19, 20]. The application of nanotechnology to medicine was shown experimentally by Lee *et al.* [21, 22] through the single red blood cell (RBC) deformability test performed by using optical trapping plastic in a microfluidic chip and through lab-on-a-chip for the transportation of RBCs in a capillary network to circulate oxygen and carbon dioxide throughout the human body [23]. Suwanpayak *et al.* [24] reported that optical trapping can be used to manipulate molecules in a liquid core waveguide and can be applied for drug delivery, wherein a PANDA ring resonator is used to form, transmit, and receive the microscopic volumes (of drugs) by controlling the ring parameters. The microscopic volume can be trapped and moved (transported) dynamically within the wavelength router or network.

Recently, a promising technique of microscopic volume trapping and transportation within an add/drop multiplexer have been reported in theory [25] and experiment [26]. Here the transporter is known as an optical tweezer. The optical tweezer generation technique is used as a powerful tool to manipulate micrometer-sized particles. To date, useful static tweezers have been well recognized and realized. Moreover, the use of dynamic tweezers has now also been realized in practical work [27–29]. Schulz *et al.* [30] have shown that it is possible to transfer trapped atoms between two optical potentials. Optical tweezers use forces exerted by intensity gradients in strongly focused beams of light to trap and move nanoscopic volumes of matter. During the movement, another combination of forces is induced between photons due to their interactions caused by photon scattering effects. The field intensity can also be adjusted and tuned to the desired gradient field and scattering force so that the suitable trapping force can be formed. Thus, by configuring the appropriated force for the transmitter/receiver part, nanoscopic particles can be transported over a long distance.

In this chapter, the methods for generating the dynamic optical tweezers/vortices using a dark soliton, a bright soliton and a Gaussian pulse propagating within an add/drop optical multiplexer with two nanoring resonators (PANDA ring resonator) will be discussed. The dynamic behavior of soliton and Gaussian pulses is well described by some authors [22]. By using the proposed system, the blood waste and unwanted substances can be trapped and transported (filtered) from the artificial human kidney. The required trapping tool sizes can be generated and formed for the specific blood waste molecules, and clean blood can be obtained and sent to its destination via the through port. However, several sensors are required for environmental and blood quality control, which is the topic for future research.

10.2 Theoretical Background

Trapping forces are exerted by the intensity gradients in the strongly focused beams of light, which can be used to trap and move microscopic volumes of matter, in which the optical forces are customarily defined by the relationship [21].

$$F = \frac{Q n_m P}{c} \tag{10.1}$$

Here Q is a dimensionless efficiency, n_m the index of refraction of the suspending medium, c the speed of light, and P the incident laser power for the specimen. Q represents the fraction of power utilized to exert force. For an incident plane wave on a perfectly absorbing particle, Q is equal to 1. To achieve stable trapping, the radiation pressure must create a stable, three-dimensional equilibrium. Because biological specimens are usually contained in aqueous medium, the dependence of F on n_m can rarely be exploited to achieve higher trapping forces. Increasing the laser power is possible, but only over a limited range due to the possibility of optical damage. Q itself is, therefore, the main determinant of the trapping force. It depends upon the numerical aperture (NA), the laser wavelength, the light polarization state, the laser mode structure, the relative index of refraction, and the geometry of the particle.

Furthermore, in the Rayleigh regime, trapping forces decompose naturally into two components. Since, in this limit, the electromagnetic field is uniform across the dielectric, particles can be treated as induced point dipoles. The scattering force is given by the following equation [21]:

$$F_{scatt} = n_m \frac{\langle S \rangle \sigma}{c} \tag{10.2}$$

where

$$\sigma = \frac{8}{3}\pi(kr)^4 r^2 \left(\frac{m^2-1}{m^2+2}\right)^2 \tag{10.3}$$

Here σ is the scattering cross-section of a Rayleigh sphere with radius r. $\langle S \rangle$ is the time averaged Poynting vector, n the index of refraction of the particle, $m = n/n_m$ the relative index, and $k = 2\pi n_m/\lambda$ is the wave number of the light. The scattering force is proportional to the energy flux and points along the direction of propagation of the incident light. The gradient field (F_{grad}) is the Lorentz force acting on the dipole induced by the light field. It is given by the following equation [21]:

$$F_{grad} = \frac{\alpha}{2}\nabla\langle E^2 \rangle \tag{10.4}$$

where

$$\alpha = n_m^2 r^3 \left(\frac{m^2-1}{m^2+2}\right) \tag{10.5}$$

The polarizability of the particle is denoted by α. The gradient force is proportional and parallel to the gradient in energy density (for $m > 1$). The large gradient force is formed by the large depth of the laser beam, in which the stable trapping requires the gradient force to be in the $-\hat{z}$ direction, which is against the direction of the incident light (dark soliton valley) and is greater than the scattering force. By increasing the NA and decreasing the focal spot size, the gradient strength can be increased [31], within a tiny system, such as a nanoscale device (nanoring resonator).

In our proposal, the trapping force is created by using a dark soliton, in which a valley of the dark soliton is generated and controlled within the PANDA ring resonator by the control port signals. Optical tweezers can be trapped, transported, and stored within a PANDA ring resonator and a wavelength router, which can be used to transport microscopic volumes of drugs (molecules) via the waveguide [32]. The trapped microscopic volumes can be manipulated within the optical tweezers [25].

10.3 Kidney On-Chip Manipulation

Microfluidics is a field burgeoning with important applications in areas such as medical devices, biotechnology, chemical synthesis, and analytical chemistry [32]. Erickson lab [33–38] has research interests that revolve around the study of micro-nanofluidics and its combination with optics to enhance flow and delivery of substances in living organisms and in implanting devices in their bodies [39–41]. In this chapter, we will propose optical trapping tools for blood manipulation dialysis (kidney dialysis) by using the PANDA ring resonator system incorporating a wavelength router. The blood cleaner on-chip is shown in Fig. 10.1. In this proposal, optical trapping and transportation technique has been used for the for blood dialysis system. In this design, the blood flows [42, 43] in through the input port of a PANDA ring resonator, and the PANDA ring resonator system 1 (P1) works in the same manner as PANDA ring resonator system 2 (P2). The whole blood can be trapped and delivered in a liquid core waveguide [44], of high and low refractive indices, depending on the blood concentration. The refractive indices for whole blood concentrations 20%, 40%, and 60% are 1.35, 1.35, and 1.37, respectively [45, 46].

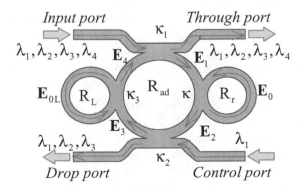

Figure 10.1   A schematic diagram of a proposed PANDA ring resonator.

When blood flows in a waveguide channel [47, 48], it is detected by a blood concentrate detector (versatile sensor, 3) [49, 50]. The most important criterion for blood dialysis is that the blood protein should be reabsorbed. To check this, a protein sensor is placed in the renal artery [51, 52] to check the protein quality before releasing into the urine. The protein sensor will detect the blood protein component and the control part will control/mitigate the whole blood circulation until it is clean. Additionally, oxygen can be fed into the system via the add port (control) to enhance the quality of whole blood and the blood pressure is monitored with the help of a wireless device [53] before the blood is delivered into the renal artery. In this case, the components of the filtrate are sodium (~0.4 nm), chlorine (~0.6 nm), and glucose (~0.72 nm) [53]. The artificial kidney is integrated in the form of a single chip, which is equivalent to one Bowman's capsule consisting of the glomerulus that is made up of capillary vessels.

A schematic diagram of a blood cleaner is shown in Fig. 10.2, which is made up of PANDA ring resonators and liquid core waveguide as shown in Fig. 10.3a. The waveguides are constructed by using different sizes and refractive indices as shown in Fig. 10.3b, using the blood cells' refractive index as 1.35. Waveguides of sizes 5–8 µm and 1 nm are formed for blood cells and proteins, respectively. The whole blood is input into the system via liquid core 1 (input port) and the required molecules are trapped and filtered by liquid core 2. Finally, the unwanted substances are trapped and filtered via the drop ports. In Fig. 10.2a, the whole blood cells are input into the blood cleaner via the input ports (E_{in}), which is equivalent to the blood flow into the afferent artery as shown in Fig. 10.2b. After the blood wastes are filtered by the glomerulus and the filtrate via the proximal tube, the cleaned blood is received via the efferent artery. In manipulation, the blood artery is replaced by the blood waveguide structure, which is formed by the liquid core waveguide as shown in Fig. 10.3. The red blood cells (RBCs), which are 5–8 µm in size, are input into the system via the input ports (liquid core 1), while unwanted substances are trapped by optical tweezers and filtered by the drop ports. Liquid core 2 is used for trapped proteins. The size of the protein molecule is 1 nm, which is smaller than the size of the RBC; therefore, the RBCs cannot transport via liquid core 2, which means that only the unwanted substances will be trapped and transported through liquid core 2.

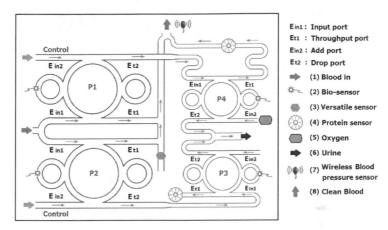

Figure 10.2 A schematic diagram of an artificial kidney manipulation on-chip system.

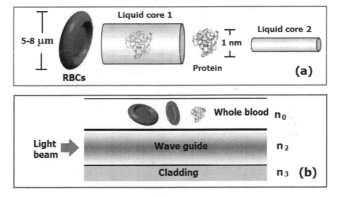

Figure 10.3 A schematic diagram of (a) blood waveguide and (b) blood waveguide structure.

To form the optical trapping tools, a bright soliton with center wavelength at 1.50 μm, peak power 4 W, and pulse 35 fs is input into the system via the input port. The coupling coefficients are given as $\kappa_0 = 0.5$, $\kappa_1 = 0.35$, $\kappa_2 = 0.1$, and $\kappa_3 = 0.35$, respectively. The ring radii are $R_{add} = 100$ μm and 75 μm, $R_R = 40$ μm and 20 μm, and $R_L = 40$ μm and 20 μm, respectively. A device with a radius of 30 nm has been reported by some authors [22]. A_{eff} are 300 μm² ($r \approx 9.77$ μm) and 0.5 μm² ($r \approx 400$ nm). In this case, the dynamic tweezers (gradient fields) for trapping the required blood waste can be in the form of bright solitons, Gaussian pulses, and dark solitons.

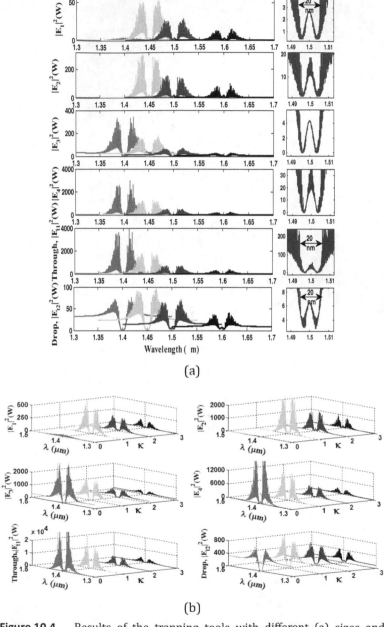

Figure 10.4 Results of the trapping tools with different (a) sizes and wavelengths and (b) tunable tweezers by coupling constant variation. R_{ad} = 100 μm and R_R = R_L = 40 μm.

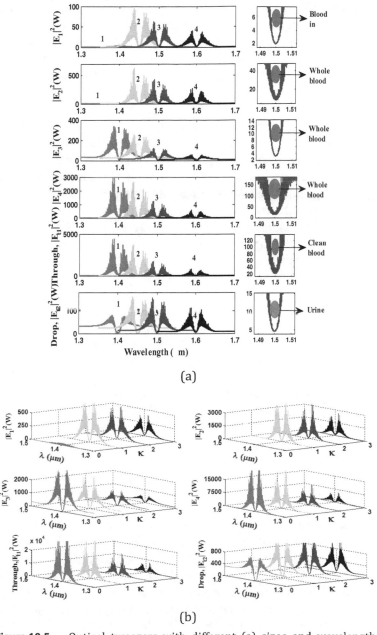

Figure 10.5 Optical tweezers with different (a) sizes and wavelengths and (b) tunable tweezers with varying coupling constants. R_{add} = 75 μm and R_R = R_L = 20 μm.

In Fig. 10.4, there are four different center wavelengths of the tweezers generated, where the dynamical movements are (a) $|E_1|^2$, (b) $|E_2|^2$, (c) $|E_3|^2$, (d) $|E_4|^2$, (e) through port signal and (f) drop port signal. In this case, the trapped molecules (blood wastes) are filtered and obtained at the drop port. By using the whole blood index, $n = 1.35$, trapping probe widths of 20 nm at the wavelength center at 1.60 μm are generated (see Fig. 10.4a). More results in terms of coupling constants and the tunable trapping obtained are shown in Fig. 10.4b. In practice, the more reliable device (blood cleaner on-chip) fabrication parameters are the ring resonator radii instead of the coupling constants. More results of the optical trapping probes generated within the PANDA ring with different wavelengths are shown in Fig. 10.5a. In this case, a bright soliton has been used as the control signal for the tunable result. Output optical tweezers of the through and drop ports with different coupling constants are shown in Fig. 10.5b, which form the basis for selecting blood-waste molecules. The clean blood can be transported into the human body via the through port. The filtering molecules (urine) are received via the drop port. The advantage of the proposed system is that the on-chip system can be fabricated on-chip and can, alternatively, be operated by a single device [54]. The proposed system can be used as a hemofiltration device, which is 90 μm deep, 200 μm wide, and 300 μm long. Unwanted substances can be trapped (filtered by the optical gradient field, from where they can be dropped by the control port (different wavelengths) [55]. Blood concentration does not affect the optical trapping ability, but since the blood refractive index is slightly different, it may affect the filtration speed.

10.4 Conclusion

In this chapter, we showed how blood waste molecules can be trapped and filtered (transported) on-chip with the help of an artificial blood cleaner (kidney). By using the appropriate dark soliton input power, the required trapping tool sizes can be controlled and obtained and, finally, clean blood can be obtained before sending it to the final destination (through port). In addition, oxygen can be fed into the system via the control port and can be used in the blood cleaning process. However, the blood cleaner on-chip system is required to work in a specific environment in which several sensors are present as shown in Fig. 10.1. Sensors, such as biosensor, blood sensor, versatile sensor, protein sensor, and blood pressure sensor, need to be integrated in the system for different purposes. The other problem is that the generated trapping tool size may cause problems in case of large molecules. Therefore, new guide pipes, mediums such as [56] nanotube materials, and sensors are the topics for future research.

References

1. L. G. Wesson, "Physiology of the human kidney," Grune & Stratton, New York (1969).

2. B. J. McNamara, B. Diouf, R.N. Douglas-Denton, M. D. Hughson, W. E. Hoy, and J. F. Bertram, "Comparison of nephron number, glomerular volume and kidney weight in Senegalese Africans and African Americans," *Nephrol. Dial. Transplant.*, **25**, 1514–1520 (2010).

3. A. Manalich, L. Reyes, M. Herrera, C. Melendi, and I. Fundora, "Relationship between weight at birth and the number and size of renal glomeruli in humans: A histomorphometric study," *Kidney Int.*, **58**, 770–777 (2000).

4. S. Rashad, and M. D. Barsoum, "Chronic kidney disease in the developing world," *N. Engl. J. Med.* **354**(1), 997–999 (2006).

5. R. Scarpioni, M. Ricardi, L. Melfa, and L. Cristinelli, "Dyslipidemia in chronic kidney disease: Are statins still indicated in reduction cardiovascular risk in patients on dialysis treatment?" *Cardiovasc. Ther.*, **28**, 361–368 (2010).

6. C. Coirault, J. C. Pourny, F. Lambert, and Y. Lecarpentier, "Mechanical property analysis of stored red blood cell using optical tweezers," *Med. Sci.*, **19**(3), 364–367 (2003).

7. S. R. Grenier, M. T. Wei, J. J. Bai, and A. Chiou, "Dynamic deformation of red blood cell in dual-trap optical tweezers," *Opt. Express*, **18**(10), 10462–10472 (2010).

8. A. R. Nissenson, C. Ronco, G. Pergamit, M. Edelstein, and R. Watts, "Continuously functioning artificial nephron system: The promise of nanotechnology," *Hemodial. Int.*, **9**, 210–217(2005).

9. F. Locatelli, F. Valderrabano, N. Hoenich, J. Bommer, K. Leunissen, and V. Cambi, "Progress in dialysis technology: Membrane selection and patient outcome," *Nephrol. Dial. Transplant.*, **15**, 1133–1139 (2000).

10. V. Gura, A. S. Macy, M. Beizai, C. Ezon, and T. A. Golper, "Technical breakthroughs in the wearable artificial kidney," *Clin. J. Am. Soc. Nephrol.*, **4**, 1441–1448 (2009).

11. L. G. Forni, and P. J. Hilton, "Continuous hemofiltration in the treatment of acute renal failure," *N. Engl. J. Med.*, **336**(18), 1303–1339 (1997).

12. M. D. Hardy, "Implantable artificial kidney," US patent 5092886. March 3, 1992.

13. W. Fissell, A. J. Fleischman, H. D. Humes, and S. Roy, "Development of continuous implantable renal replacement: Past and future," *Transl. Res.*, **150**(6), 327–336 (2007).

14. T. M. Dykstra, "Kidney dialysis method and device," US patent 5876366. March 2, 1999.

15. S. J. Lu, Q. Feng, J. S. Park, *et al.*, "Biological properties and enucleation of red blood cells from human embryonic stem cells, *Blood*, **112**, 4362–4363 (2008).

16. A. Ashkin, J. M. Dziedzic, and T. Yamane, "Observation of a single-beam gradient force optical trap for dielectric," *Opt. Lett.*, **11**, 288–290 (1986).

17. H. D. Chen, K. Ge, Y. Li, *et al.*, "Application of optical tweezers in the research of molecular interaction between lymphocyte function associated antigen-1 and its monoclonal antibody," *Cell Mol. Immunol.*, **4**, 221–225 (2007).

18. W. G. Lee, K. Park, H. Bang, *et al.*, "Single red blood cell deformity test using optical trapping in plastic microfluid chip," *Proceedings of the 31 Annual International IEEE EMBS Special Topic on Conference Microtechnologies in Medicine and Biology Kahuku, Oahu, Hawaii, 12–15 May, 2005*, pp. 389–390.

19. D. Obrist, B. Weber, A. Buck, and P. Jenny, "Red blood cell distribution in simplified capillary," *Phil. Trans. R. Soc. A.*, **368**, 2897–2918 (2010).

20. Y. C. Chen, G. Y. Chen, Y. C. Lin, and G. J. Wang, "A lab-on-a-chip capillary network for red blood cell hydrodynamics," *Microfluid. Nanofluid.*, **9**, 585–591 (2010).

21. N. Suwanpayak, M. A. Jalil, C. Teeka, J. Ali, and P. P. Yupapin, "Optical vortices generated by a PANDA ring resonator for drug trapping and delivery applications," *Bio. Med. Opt. Express*, **2**(1), 159–168 (2011).

22. B. Piyatamrong, K. Kulsirirat, W. Techithdeera, S. Mitatha, and P. P. Yupapin, "Dynamic potential well generation and control using double resonators incorporating in an add/drop filter," *Mod. Phys. Lett. B*, **24**, 3071–3082 (2010).

23. H. Cai, and A. Poon, "Optical manipulation and transport of microparticle on silicon nitride microring-resonator-based add-drop devices," *Opt. Lett.*, **35**, 2855–2857 (2010).

24. A. Ashkin, J. M. Dziedzic, and T. Yamane, "Optical trapping and manipulation of single cells using infrared laser beams," *Nature*, **330**, 769–771 (1987).

25. K. Egashira, A. Terasaki, and T. Kondow, "Photon-trap spectroscopy applied to molecules adsorbed on a solid surface: Probing with a standing wave versus a propagating wave," *Appl. Opt.*, **80**, 5113–5115 (1998).

26. A. V. Kachynski, A. N. Kuzmin, H. E. Pudavar, D. S. Kaputa, A. N. Cartwright, and P. N. Prasad, "Measurement of optical trapping forces by use of the two-photon-excited fluorescence of microspheres," *Opt. Lett.*, **28**, 2288–2290 (2003).

27. M. Schulz, H. Crepaz, F. Schmidt-Kaler, J. Eschner, and R. Blatt, "Transfer of trapped atoms between two optical tweezer potentials," *J. Mod. Opt.*, **54**, 1619–1626 (2007).

28. M. Tasakorn, C. Teeka, R. Jomtarak, and P. P. Yupapin, "Multitweezers generation control within a nanoring resonator system," *Opt. Eng.*, **49**, 075002 (2010).

29. K. Svoboda, and S. M. Block, "Biological applications of optical forces," *Annu. Rev. Biophys. Biomol. Struct.*, **23**, 247–283 (1994).

30. J. Zhu, S. K. Ozdemir, Y. F. Xiao, *et al.*, "On-chip single nanoparticle detection and sizing by mode splitting in an ultrahigh-Q micro-resonator," *Nat. Photonics*, **4**, 46–49 (2010).

31. N. Suwanpayak, and P. P. Yupapin, "Molecular buffer using a PANDA ring resonator for drug delivery use," *Int. J. Nanomedicine*, **6**, 575–580 (2011).

32. D. Psaltis, S. R. Quake, and C. Yang, "Developing optofluidic technology through the fusion of microfluidics and optics," *Nature*, **442**(27), 381–386 (2006).

33. A. J. H. Yang, and D. Erickson, "Optofluidic ring resonator switch for optical particle transport," *Lab Chip*, **10**, 769–774 (2010).

34. A. J. Chung, Y. S. Huh, and D. Erickson, "A robust, electrochemically driven microwell drug delivery system for controlled vasopressin release," *Biomed. Microdevices*, **11**, 861–867 (2009).

35. M. Krishnan, M. Tolley, H. Lipson, and D. Erickson D, "Hydrodynamically tunable affinities for fluidic assembly," *Langmuir*, **25**, 3769–3744 (2009).

36. A. J. Chung, and D. Erickson, "Engineering insect flight metabolics using immature stage implanted microfluidics," *Lab Chip*, **9**, 669–676 (2009).

37. B. S. Schmidt, A. H. J. Yang, D. Erickson, and M. Lipson, "Optofluidic trapping and transport on solid core waveguides within a microfluidic device," *Opt. Express*, **1**(22), 14322–14334 (2007).

38. M. Segev, D. N. Christodoulides, and C. Rotschild, "Method and system for manipulating fluid medium," US patent 2011/0023973 A. February 3, 2011.

39. M. Bugge, and G. Palmers, "Implantable device for utilization of the hydraulic energy of the heart," US patent RE41394 E. June 22, 2010.

40. S. Y. Chen, S.H. Hu, D. M. Liu, and K. T. Kuo, "Drug delivery nanodevice, its preparation method and used thereof," US patent 2011/0014296 A1. January 20, 2011.

41. K. I. Yamamoto, K. Kobayashi, K. Endo, *et al.*, "Hollow-fiber blood-dialysis membranes: Superoxide generation, permeation, and dismutation measured by chemiluminescence," *J. Artif. Organs*, **8**, 257–262 (2005).

42. H. W. Huang, T. C. Ching Shih, and C. T. Liauh, "Predicting effects of blood flow rate and size of vessels in a vasculature on hyperthermia treatments using computer simulation," *Bio. Med. Eng. OnLine*, **9**(1), 18 (2010).

43. A. G. Hudetz, "Blood flow in the cerebral capillary network: A review emphasizing observations with intravital microscopy," *Micro-circulation*, **4**(2), 233–252 (1997).

44. D. K. Sardar, and L. B. Levy, "Optical properties of whole blood," *Laser. Med. Sci.*, **13**, 106–111 (1998).

45. L. M. Bonanno, and L. A. DeLouise, "Whole blood optical biosensor," *Biosens. Bioelectron.*, **23**, 444–448 (2007).

46. R. V. Harrison, N. Harel, J. Panesar, and R. J. Mount, "Blood capillary distribution correlates with hemodynamic-based functional imaging in cerebral cortex," *Cereb. Cortex*, **12**, 225–233 (2002).

47. E. Turkstra, B. Braam, and H. A. Koomans, "Impaired renal blood flow autoregulation in two-kidney, one-clip hypertensive rats is caused by enhanced activity of nitric oxide," *J. Am. Soc. Nephrol.*, **11**, 847–855 (2000).

48. R. J. Thomson, "Blood concentrate detector," US patent 2009/0310123 A1. December 17, 2009.

49. S. Carrara, "Nano-bio-technology and sensing chips: New systems for detection in personalized therapies and cell biology," *Sensors*, **10**, 526–543 (2010).

50. P. V. Preejith, C. S. Lim, A. Kishen, M. S. John, and A. Asundi, "Total protein measurement using a fiber-optic evanescent wave-based biosensor," *Biotechnol. Lett.*, **25**, 105–110 (2003).

51. J. Das, and S. O. Kelley, "Protein detection using arrayed microsensor chips: Tuning sensor footprint to achieve ultrasensitive readout of CA-125 in serum and whole blood," *Anal. Chem.*, **83**(4), 1167–1172 (2011).

52. K. Oyri, I. Balasingham, E. Samset, J. O. Høgetveit, and E. Fosse, "Wireless continuous arterial blood pressure monitoring during surgery: A pilot study," *Anesth. Analg.*, **102**, 478–483 (2006).

53. T. W. Hsu, Y. C. Chen, M. J. Wu, A. F. Y. Li, W. C. Yang, and Y. Y. Ng, "Reinfusion of ascites during hemodialysis as a treatment of massive refractory ascites and acute renal failure," *Int. J. Nephrol. Renovasc. Dis.*, **4**, 29–33 (2011).

54. J. T. Santini, A. C. Richards, R. Scheidt, M. J. Cima, R. Langer, "Microchip as controlled drug delivery devices," *Angew. Chem. Int. Ed.*, **39**, 2396–2407 (2000).

55. A. Jesacher, C. Maurer, S. Furhapter, A. Schwaighofer, S. Bernet, and M. Ritsch-Marte, "Optical tweezers of programmable shape with transverse scattering forces," *Opt. Commun.*, **281**, 2207–2212 (2008).

56. A. Ashkin, "Optical trapping and manipulation of neutral particles using lasers," *Proc. Natl. Acad. Sci.*, **94**, 4853–4860 (1997).

Chapter 11

All-Optical Logic XOR/XNOR Gates

11.1 Introduction

To date, many researchers have demonstrated interesting techniques that can be used to realize the various optical logic functions
(i.e., AND, NAND, OR, XOR, XNOR, NOR) by using different schemes,
including thermo-optic effect in two cascaded microring resonators
[1], quantum dot [2, 3], semiconductor optical amplifier (SOA)
[4–11], TOAD-based interferometer device [12], nonlinear effects in
SOI waveguide [13, 14], nonlinear loop mirror [15, 16], DPSK format
[17, 18], local nonlinear in MZI [19], photonic crystal [20, 21], error
correction in multipath differential demodulation [22], fiber optical
parametric amplifier [23], multimode interference in SiGe/Si [24],
polarization and optical processor [25], and injection-locking effect
in semiconductor laser [26]. Although the search for new techniques
is ongoing, there is some room for techniques that can be good
candidates for optical logic gates. Therefore, in this work, we propose
the use of the simultaneous arbitrary two-input logic XOR/XNOR and
all-optical logic gates based on dark-bright soliton conversion within
the add/drop optical filter system. The advantage of the scheme is
that random codes can be generated simultaneously by using the
dark-bright soliton conversion behavior, in which the coincident
dark and bright solitons can be separated after propagating them
into a $\pi/2$ phase retarder, which can be used to form the security
codes. Moreover, this is a simple and flexible scheme for an arbitrary
logic switching system that can be used to form advanced complex

Nanophotonics: Devices, Circuits, and Systems
Preecha P. Yupapin, Keerayoot Srinuanjan, and Surachart Kamoldilok
Copyright © 2013 Pan Stanford Publishing Pte. Ltd.
ISBN 978-981-4364-36-2 (Hardcover), 978-981-4364-37-9 (eBook)
www.panstanford.com

logic circuits. The proposed scheme is based on a one-bit binary comparison XOR/XNOR scheme that can be compared to any two bits, i.e., between 0 and 0 (dark-dark solitons), 0 and 1 (dark-bright solitons), 1 and 0 (bright-dark solitons), or 1 and 1 (bright-bright solitons). This which will be detailed in the following section.

11.2   Dark-Bright Soliton Conversion Mechanism

In operation, dark-bright soliton conversion that uses a ring resonator optical channel dropping filter (OCDF) is composed of two sets of coupled waveguides as shown in Fig. 11.1a, b, where for convenience, Fig. 11.1b is the equivalence diagram of Fig.11.1a. The relative phase of the two output light signals after coupling into the optical coupler and before coupling into the ring and the input bus is $\partial/2$. This means that the signals that couple into the drop and through ports acquired a phase of π with respect to the input port signal. If coupling coefficients are engineered appropriately, then the fields coupled into the through port would completely extinguish the wavelength on resonance and all power would be coupled into the drop port. We will show how this is possible later in this section.

$$E_{ra} = -j\kappa_1 E_i + \tau_1 E_{rd} \tag{11.1}$$

$$E_{rb} = \exp(j\omega T/2)\exp(-\alpha L/4)E_{ra} \tag{11.2}$$

$$E_{rc} = \tau_2 E_{rb} - j\kappa_2 E_a \tag{11.3}$$

$$E_{rd} = \exp(j\omega T/2)\exp(-\alpha L/4)E_{rc} \tag{11.4}$$

$$E_t = \tau_1 E_i - j\kappa_1 E_{rd} \tag{11.5}$$

$$E_d = \tau_2 E_a - j\kappa_2 E_{rb} \tag{11.6}$$

Here E_i is the input field, E_a the add (control) field, E_t the through field, E_d the drop field, $E_{ra}...E_{rd}$ the fields in the ring at points a...d, κ_1 the field coupling coefficient between the input bus and ring, κ_2 the field coupling coefficient between the ring and output bus, L s the circumference of the ring, T the time taken for one roundtrip (roundtrip time), and α the power loss in the ring per unit length. We assume that this is lossless coupling, i.e., $\tau_{1,2} = \sqrt{1-\kappa_{1,2}^2}$, and $T = Ln_{\text{eff}}/c$.

The output power/intensities at the drop and through ports are given by

$$|E_d|^2 = \left| \frac{-\kappa_1 \kappa_2 A_{1/2} \Phi_{1/2}}{1 - \tau_1 \tau_2 A\Phi} E_i + \frac{\tau_2 - \tau_1 A\Phi}{1 - \tau_1 \tau_2 A\Phi} E_a \right|^2 \tag{11.7}$$

$$|E_t|^2 = \left| \frac{\tau_2 - \tau_1 A\Phi}{1 - \tau_1 \tau_2 A\Phi} E_i + \frac{-\kappa_1 \kappa_2 A_{1/2} \Phi_{1/2}}{1 - \tau_1 \tau_2 A\Phi} E_a \right|^2 \tag{11.8}$$

Here $A_{1/2} = \exp(-\alpha L/4)$ (the half-roundtrip amplitude), $A = A_{1/2}^2$, $\Phi_{1/2} = \exp(j\omega T/2)$ (the half-roundtrip phase contribution), and $\Phi = \Phi_{1/2}^2$.

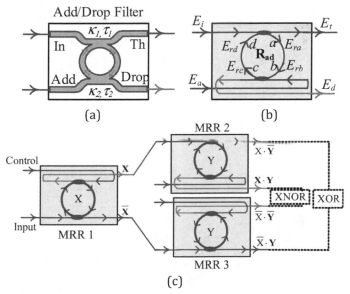

Figure 11.1   A schematic diagram of a simultaneous optical logic XOR and XNOR gate.

The input and control fields at the input and add ports are formed by the dark-bright optical soliton, expressed by Eqs. 11.9 and 11.10.

$$E_{in}(t) = E_0 \tanh\left[\frac{T}{T_0}\right] \exp\left[\left(\frac{z}{2L_D}\right) - i\omega_0 t\right] \tag{11.9}$$

$$E_{in}(t) = E_0 \operatorname{sech}\left[\frac{T}{T_0}\right] \exp\left[\left(\frac{z}{2L_D}\right) - i\omega_0 t\right] \tag{11.10}$$

Here E_0 and z are the optical field amplitude and propagation distance, respectively. $T = t - \beta_1 z$, where β_1 and β_2 are the coefficients of the linear and second-order terms of Taylor expansion of the propagation constant. $L_D = T_0^2/|\beta_2|$ is the dispersion length of the soliton pulse. T_0 in equation is a soliton pulse propagation time at initial input (or the soliton pulse width), where t is the soliton phase-shift time and ω_0 is the frequency shift of the soliton. When the optical field is entered into the nanoring resonator as shown in Fig. 11.2, where the coupling coefficient ratio, $\kappa_1:\kappa_2$, is 50:50, 90:10, or 10:90, a dark soliton is input into the input and control ports as shown in Fig. 11.1 and the results obtained are shown in Fig. 11.2a. The results for dark and bright solitons used as input and control signals are shown in Fig 11.2b, for bright and dark solitons in Fig. 11.2c, and for a bright soliton in Fig. 11.2d. The ring radii R_{ad} = 5 μm, A_{eff} = 0.25 μm², n_{eff} = 3.14 (for InGaAsP/InP), α = 5 dB/mm, γ = 0.1, and λ_0 = 1.51 μm.

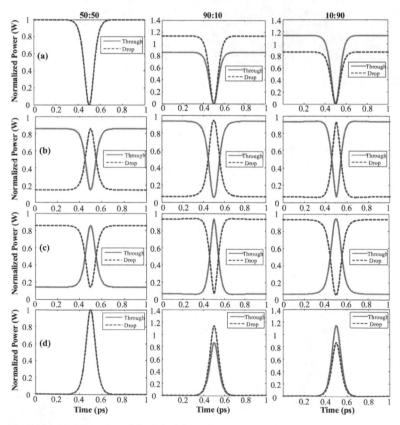

Figure 11.2   Results of dark-bright soliton conversion.

11.3   Optical XOR/XNOR Logic Gate Operation

The proposed architecture is schematically shown in Fig. 11.1c. A continuous optical wave with a wavelength λ is formed by an optical dark-bright soliton pulse train X by using MRR 1, in which the optical pulse trains that appear at the through and drop ports of MRR 1 are $\overline{X}$ and X, respectively ($\overline{X}$ is the inverse of X or dark-bright conversion). It is assumed that the input optical dark-bright soliton wave is directed to the drop port when the optical signal is 1 (dark soliton pulse). In other words, the MRR 1 resonates at λ when the input dark soliton pulse is applied.

If the optical pulse train X is fed into MRR 2 solely from its input port and is formed by an optical pulse train Y bit by bit using MRR 2, assuming that no signal is fed into MRR 2 from its add port, then the optical pulse trains that appear at the through and drop ports of MRR 2 will be $X \cdot \overline{Y}$ and $X \cdot Y$, respectively. The symbol represents the logical operation AND here.

If the optical pulse train $\overline{X}$ is fed into MRR 3 from its input port solely and is formed by an optical pulse train Y bit by bit using MRR 3, assuming that no signal is fed into MRR 3 from its input port, then the optical pulse trains that appear at the through and drop ports of MRR 3 will be $\overline{X} \cdot Y$ and $\overline{X} \cdot \overline{Y}$, respectively. If the optical pulse trains X and $\overline{X}$ are fed into MRR 2 and MRR 3 from their input ports simultaneously (see Fig. 11.1c), the optical pulse trains $X \cdot \overline{Y} + \overline{X} \cdot Y$ and $X \cdot Y + \overline{X} \cdot \overline{Y}$ are achieved at the through and drop ports of MRR 2 and MRR 3, respectively (see Fig. 11.1c). The symbol + represents the logical operation OR, which is implemented through the multiplexing function of MRR 2 and MRR 3. It is well known that the XOR and XNOR operations can be calculated by using the formulas $X \oplus Y = X \cdot \overline{Y} + \overline{X} \cdot Y$ and $X \otimes Y = X \cdot Y + \overline{X} \cdot \overline{Y}$, where the capital letters represent logical variables and the symbols $\oplus$ and $\otimes$ represent the XOR and XNOR operators, respectively. Therefore, the proposed architecture can be used as an XOR and XNOR calculator.

By using the dynamic performance of the device as shown in Fig. 11.3, two pseudo-random binary sequence $2^4 - 1$ signals at 100 Gbit/s are converted to be two optical signals bit by bit according to the rule presented above and then applied to the corresponding MRRs. Clearly, a logic 1 is obtained when the applied optical bright soliton pulse signals are generated and a logic 0 is obtained when the applied optical dark soliton pulse signals are generated. Therefore, the device performs the XOR and XNOR operations correctly.

The proposed simultaneous all-optical logic XOR and XNOR gates device is shown in Fig. 11.1c. The input and control light pulse trains are input into the first add/drop optical filter (MRR 1) by using the dark solitons (logic "0") or the bright solitons (logic "1"). First, the dark soliton is converted to dark and bright solitons via the add/drop optical filter, which can be seen at the through and drop ports with π phase shift [27], respectively. By using the add/drop optical filters (MRR 2 and MRR 3), both input signals are generated by the first-stage add/drop optical filter. Next, the input data "Y" with logic "0" (dark soliton) and logic "1"(bright soliton) are added into both add ports, the dark-bright soliton conversion with π phase shift is operated again. For large scale operation (Fig. 11.1c), results obtained are simultaneously seen by D_1, D_2, T_1, and T_2 at the drop and through ports for optical logic XNOR and XOR gates, respectively.

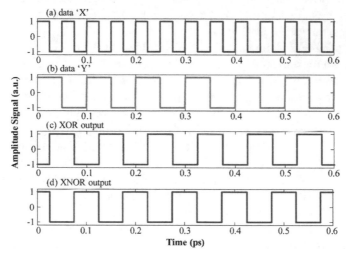

Figure 11.3 Output results of the dynamic performance of the device when (a) data "X," (b) data "Y," (c) all-optical XOR gate, and (d) XNOR logic gate.

In simulation, the add/drop optical filter parameters are fixed for all coupling coefficients to be κ_s = 0.05, R_{ad} = 5 μm, A_{eff} = 0.25 μm² [28], α = 5 dBmm⁻¹ for all add/drop optical filters in the system. Results of the simultaneous optical logic XOR and XNOR gates are generated by using dark-bright soliton conversion with wavelength center at λ_0 = 1.51 μm and pulse width 45 fs. In Fig. 11.4, simulation result of the simultaneous output optical logic gate is seen when the input data logic "00" is added, whereas the obtained output optical logic is "**0001**" (see Fig. 11.4a). Similarly, when the simultaneous output optical logic gate input data logic "01" is added, the output optical logic "**0010**" is formed (see Fig. 11.4b). Next, when the output optical logic gate input data logic "10" is added, the output optical logic "**1000**" is formed (see Fig. 11.4c). Finally, when the output optical logic input data logic "11" is added, we found that the output optical logic "**0100**" is obtained (see Fig. 11.4d). The simultaneous optical logic gate output is concluded in Table 11.1. We found that the output data logic in the drop ports D_1 and D_2 are optical logic XNOR gates whereas the output data logic in the through ports T_1 and T_2 are optical logic XOR gates, the switching time being 55.1 fs.

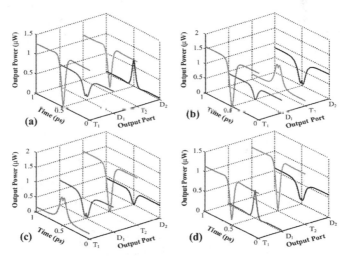

Figure 11.4   The output logic XOR/XNOR gates when the input logic states are (a) "DD," (b) "'DB," (c) "BD," and (d) "BB."

Table 11.1   Output of optical logic XOR and XNOR gates

Input data		Output logic					
		(T_1)	(D_1)	(T_2)	(D_2)	XOR	XNOR
X	Y	$X \cdot \overline{Y}$	$X \cdot Y$	$\overline{X} \cdot Y$	$\overline{X} \cdot \overline{Y}$	$X \cdot \overline{Y} + \overline{X} \cdot Y$	$X \cdot Y + \overline{X} \cdot \overline{Y}$
D	D	D	D	D	B	D	B
D	B	D	D	B	D	B	D
B	D	B	D	D	D	B	D
B	B	D	B	D	D	D	B

Note: D (dark soliton) = logic "0," and B (bright soliton) = logic "1".

11.4 Operating Principle of Simultaneous All-Optical Logic Gates

The configuration of the proposed simultaneous all-optical logic gates is shown in Fig. 11.5. The input and control light ("A") pulse trains in the first add/drop optical filter (No. "01") are the dark soliton (logic "0"). In the first stage of the add/drop filter, the dark-bright soliton conversion is seen at the through and drop ports with π phase shift. In the second stage, (no. 11 and 12), both inputs are generated by the first stage of the add/drop optical filter, in which the input data "B" with logic "0" (dark soliton) and logic "1" (bright soliton) are added into both add ports. The outputs of the second stage are dark-bright soliton conversion with π phase shift again. In the third stage of the add/drop optical filter (no. 21 to 24), the input data "C" with logic "0" (dark soliton) and logic "1" (bright soliton) are inserted into all final stage add ports. In the final stage of the add/drop optical filter (no. 31 to 38), the input data "D" with logic "0" (dark soliton) and logic "1" (bright soliton) are inserted into all final-stage add ports and the output numbers 1 to 16 are shown simultaneously at the all-optical logic gate.

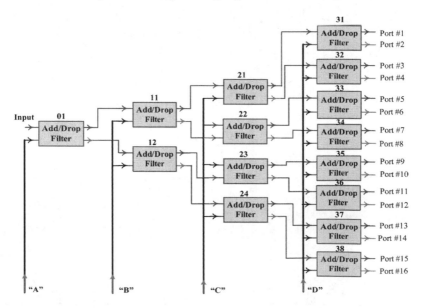

Figure 11.5   A schematic diagram of the proposed all-optical logic gate.

Table 11.2   Output results of all-optical logic gates

"A"	Output Port No.															
	1	2	3	4	5	6	7	8	9	10	11	12	13	14	15	16
	0	0	0	0	0	0	0	0	0	0	0	0	0	0	0	0
	0	1	0	1	0	1	0	1	0	1	0	1	0	1	0	1
	0	0	1	0	0	0	1	0	0	0	1	0	0	0	1	0
DARK ("0")	0	1	1	1	0	1	1	1	0	1	1	1	0	1	1	1
	0	0	0	0	1	0	0	0	0	0	0	0	1	0	0	0
	0	1	0	1	1	1	0	1	0	1	0	1	1	1	0	1
	0	0	1	0	1	0	1	0	0	0	1	0	1	0	1	0
	0	1	1	1	1	1	1	1	0	1	1	1	1	1	1	1
	0	0	0	0	0	0	0	0	1	0	0	0	0	0	0	0
	0	1	0	1	0	1	0	1	1	1	0	1	0	1	0	1
	0	0	1	0	0	0	1	0	1	0	1	0	0	0	1	0
BRIGHT ("1")	0	1	1	1	0	1	1	1	1	1	1	1	0	1	1	1
	0	0	0	0	1	0	0	0	1	0	0	0	1	0	0	0
	0	1	0	1	1	1	0	1	1	1	0	1	1	1	0	1
	0	0	1	1	1	0	1	0	1	0	1	0	1	0	1	0
	0	1	1	1	1	1	1	1	1	1	1	1	1	1	1	1

⬛ AND, ⬛ XNOR, ⬛ XOR, ⬜ NAND

The simulation parameters of the add/drop optical filters are fixed for all coupling coefficients as $\kappa_s = 0.05$, $R_{ad} = 5$ μm, $A_{eff} = 0.25$ μm^2 [27, 28], and $\alpha = 5$ dBmm^{-1} for all add/drop optical filter system. Simulation results of the simultaneous all-optical logic gates are generated by using the dark-bright soliton conversion at the wavelength center $\lambda_0 = 1.51$ μm and pulse width 45 fs. Fig. 11.6

shows the simulation result of simultaneous output optical logic gate when the input and control signals ("A") are dark solitons. The input data "BCD" are (a) "DDD," (b) "DDB," (c) "DBD," (d) "DBB," (e) "BDD," (f) "BDB," (g) "BBD" and (h) "BBB." Results of all outputs for all-optical logic gates are concluded in Table 11.2.

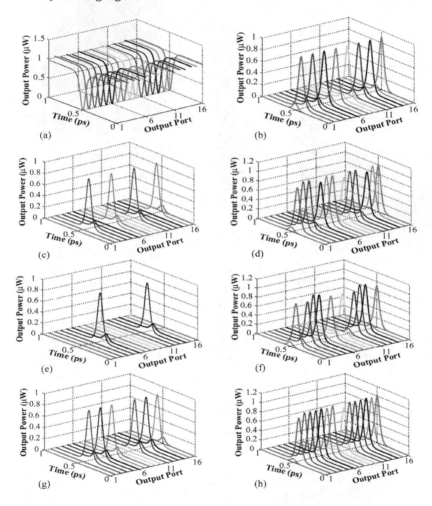

(a) (b) (c) (d) (e) (f) (g) (h)

Figure 11.6 Simulation results of all-optical logic gate when the input and control signals ("A") are dark soliton pulses. The inputs "BCD" are (a) "DDD," (b) "DDB," (c) "DBD," (d) "DBB," (e) "BDD," (f) "BDB," (g) "BBD," and (h) "BBB."

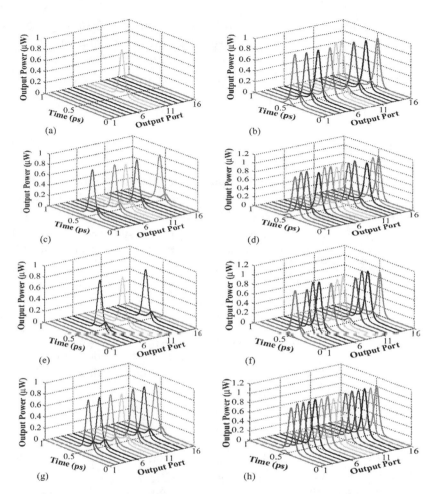

Figure 11.7 Simulation results of an all-optical logic gate when the dark soliton input and control signals ("A") are bright soliton pulses. The inputs "BCD" are (a) "DDD," (b) "DDB," (c) "DBD," (d) "DBB," (e) "BDD," (f) "BDB," (g) "BBD," and (h) "BBB."

11.5 Conclusion

In this chapter, we proposed a novel technique that can be used to generate optical logic XOR/XNOR gates by using dark-bright soliton conversion within an add/drop optical filter system. On using the dark-bright soliton conversion concept, the results showed that the input logic "0" and the control logic "0" can be formed by using the dark soliton (D) pulse trains. We also found that the simultaneous optical logic XOR/XNOR can be seen randomly both at the drop and through ports, which shows that this technique has the potential of being applied for large-scale use, especially in security code applications.

References

1. L. Zhang, R. Ji, L. Jia, *et al.*, "Demonstration of directed XOR/XNOR logic gates using two cascaded microring resonators," *Opt. Lett.*, **35**(10), 1620–1622(2010).

2. S. Ma, Z. Chen, H. Sun, and K. Dutta, "High speed all-optical logic gates based on quantum dot semiconductor optical amplifiers," *Opt. Express*, **18**(7), 6417–6422 (2010).

3. T. Kawazoe, K. Kobayashi, K. Akahane, M. Naruse, N. Yamamoto, and M. Ohtsu, "Demonstration of nanophotonic NOT gate using near-field optically coupled quantum dots," *Appl. Phys. B*, **84**, 243–246 (2006).

4. J. Dong, X. Zhang, and D. Huang, "A proposal for two-input arbitrary Boolean logic gates using single semiconductor optical amplifier by picoseconds pulse injection," *Opt. Express*, **17**(10), 7725–7730 (2009).

5. J. Dong, X. Zhang, J. Xu, and D. Huang, "40 Gb/s all-optical logic NOR and OR gates using a semiconductor optical amplifier: Experimental demonstration and theoretical analysis," *Opt. Commun.*, **281**, 1710–1715 (2008).

6. B. C. Han, J. L. Yu, W. R. Wang, L. T. Zhang, H. Hu, and E. Z. Yang, "Experimental study on all-optical half-adder based on semiconductor optical amplifier," *Optoelectron. Lett.*, **5**(3), 0162–0164 (2009).

7. Z. Li, Y. Liu, S. Zhang, *et al.*, "All-optical logic gates using semi-conductor optical amplifier assisted by optical filter," *Electron. Lett.*, **41**, 1397–1399 (2005).

8. S. H. Kim, J. H. Kim, B. G. Yu, *et al.*, "All-optical NAND gate using cross-gain modulation in semiconductor optical amplifiers," *Electron. Lett.*, **41**, 1027–1028 (2005).

9. X. Zhang, Y. Wang, J. Sun, D. Liu, and D. Huang, "All-optical AND gate at 10 Gbit/s based on cascaded single-port-couple SOAs," *Opt. Express*, **12**, 361–366 (2004).

10. Z. Li, and G. Li, "Ultrahigh-speed reconfigurable logic gates based on four-wave mixing in a semiconductor optical amplifier," *IEEE Photon. Technol. Lett.*, **18**, 1341–1343 (2006).

11. S. Kumar, and A. E. Willner, "Simultaneous four-wave mixing and cross-gain modulation for implementing an all-optical XNOR logic gate using a single SOA," *Opt. Express*, **14**, 5092–5097 (2006).

12. J. N. Roy, and D. K. Gayen, "Integrated all-optical logic and arithmetic operations with the help of a TOAD-based interferometer device-alternative approach," *Appl. Opt.*, **46**(22), 5304–5310 (2007).

13. M. Khorasaninejad, and S. S. Saini, "All-optical logic gates using nonlinear effects in silicon-on-insulator waveguides," *Appl. Opt.*, **48**(25), F31–F36 (2009).

14. Y. D. Wu, "All-optical logic gates by using multibranch waveguide structure with localized optical nonlinearity," *IEEE J. Sel. Top. Quantum Electron.*, **11**(2), 307–312 (2005).

15. Y. Miyoshi, K. Ikeda, H. Tobioka, *et al.*, "Ultrafast all-optical logic gate using a nonlinear optical loop mirror based multi-periodic transfer function," *Opt. Express*, **16**(4), 2570–2577 (2008).

16. T. Houbavlis, K. Zoiros, A. Hatziefremidis, *et al.*, "10 Gbit/s all-optical Boolean XOR with SOA fibre Sagnac gate," *Electron. Lett.*, **35**, 1650–1652 (1999).

17. J. Wang, Q. Sun, and J. Sun, "All-optical 40 Gbit/s CSRZ-DPSK logic XOR gate and format conversion using four-wave mixing," *Opt. Express*, **17**(15), 12555–12563 (2009).

18. J. Xu, X. Zhang, Y. Zhang, J. Dong, D. Liu, and D. Huang, "Reconfigurable all-optical logic gates for multi-input differential phase-shift keying signals: Design and experiments," *J. Lightwave Technol.*, **27**(23), 5268–5275 (2009).

19. Y. D. Wu, and T. T. Shih, "New all-optical logic gates based on the local nonlinear Mach-Zehnder interferometer," *Opt. Express*, **16**(1), 248–257 (2008).

20. Y. Zhang, Y. Zhang, and B. Li, "Optical switches and logic gates based on self-collimated beams in two-dimensional photonic crystals," *Opt. Express*, **15**(15), 9287–9292 (2007).

21. G. Berrettini, A. Simi, A. Malacarne, A. Bogoni, and L. Poti, "Ultrafast integrable and reconfigurable XNOR, AND, NOR, and NOT photonic logic gate," *IEEE Photon. Technol. Lett.*, **18**(8), 917–919 (2006).

22. Y. K. Lize, L. Christen, M. Nazarathy, *et al.*, "Combination of optical and electronic logic gates for error correction in multipath differential demodulation," *Opt. Express*, **15**(11), 6831–6839 (2007).

23. D. M. F. Lai, C. H. Kwok, and K. K. Y. Wong, "All-optical picoseconds logic gates based on a fiber optical parametric amplifier," *Opt. Express*, **16**(22), 18362–18370 (2008).

24. Z. Li, Z. Chen, and B. Li, "Optical pulse controlled all-optical logic gates in SiGe/Si multimode interference," *Opt. Express*, **13**(3), 1033–1038 (2005).

25. Y. A. Zaghloul, and A. R. M. Zaghloul, "Complete all-optical processing polarization-based binary logic gates and optical processors," *Opt. Express*, **14**(21), 9879–9895 (2006).

26. L. Y. Han, H. Y. Zhang, and Y. L. Guo, "All-optical NOR gate based on injection-locking effect in a semiconductor laser," *Optoelectron. Lett.*, **4**(1), 0034–0037 (2008).

27. S. Mookherjea, and M. A. Schneider, "The nonlinear microring add-drop filter," *Opt. Express*, **16**, 15130–15136 (2008).

28. Y. Kokubun, Y. Hatakeyama, M. Ogata, S. Suzuki, and N. Zaizen, "Fabrication technologies for vertically coupled microring resonator with multilevel crossing busline and ultracompact-ring radius," *IEEE J. Sel. Top. Quantum Electron.*, **11**, 4–10 (2005).

Chapter 12

Laser Gun Design

12.1 Introduction

Recently, it was pointed out that time has come for the generation of the science fiction weapon known as a death ray into a realistic tool [1]. The weapon can be used in many applications, such as nuclear weapon or missile defense, laser gun, and medical tools. In this chapter, we will show some interesting results obtained by using a laser within a tiny device. An optical soliton can be recognized as a powerful light source for a high-powered laser. However, a pumping system is required before the soliton can be generated. For simplicity, a Gaussian soliton is recommended to form the soliton instead of pumping the soliton. One interesting aspect of the Gaussian soliton is that a non-dispersive soliton can be realized by using a 1.30 μm light source. Many research works have been reported on the use of a Gaussian pulse [1–6]. Recently, an interesting aspect of the propagation of a light pulse within a nonlinear microring device has been reported [7], in which the transfer function of the output at resonant condition is derived and studied. Researchers found that the broad spectrum of a light pulse can be transformed into discrete pulses. An optical soliton is used to enlarge the optical bandwidth when a Gaussian pulse is propagating within the nonlinear microring resonator [8, 9]. The superposition of self-phase modulation (SPM) soliton pulses, either bright or dark [10] solitons, can generate a large output power. Hasegawa *et al.* has described many applications of solitons in optical fibers [11]. Many of the soliton-related concepts

Nanophotonics: Devices, Circuits, and Systems
Preecha P. Yupapin, Keerayoot Srinuanjan, and Surachart Kamoldilok
Copyright © 2013 Pan Stanford Publishing Pte. Ltd.
ISBN 978-981-4364-36-2 (Hardcover), 978-981-4364-37-9 (eBook)
www.panstanford.com

in fiber optic are discussed by Agrawal [12]. The problems of soliton-soliton interactions [13], collision [14], rectification [15], and dispersion management [16] are yet to be addressed and solved. In this chapter, we will discuss about a common laser source that can be used to generate a broad spectrum of lasers, especially, with broad center wavelengths that range from 0.40 to 1.50 μm. By using suitable microring parameters, it has been shown that optical signals, i.e., Gaussian pulses, can be amplified within a nonlinear ring resonator system. A flat laser output can be generated with sufficiently high power, which can be used in applications such as a laser gun and a laser cutting device. In addition, a compact device can be constructed by the small embedded devices, which can be used for practical purposes.

12.2   High-Power Laser Generation

For high-power laser generation, light from a monochromatic light source is launched into a nonlinear microring resonator. The combination of terms in attenuation (α) and phase (ϕ_0) constants results in the temporal coherence degradation. Hence, the time-dependent input light field (E_{in}), without pumping the term, can be expressed as

$$E_{in}(t) = E_0\, e^{-\alpha L + j\phi_0(t)} \tag{12.1}$$

Here L is the propagation distance (waveguide length).

We assume that the nonlinearity of the optical ring resonator is of Kerr type. The refractive index is given by

$$n = n_0 + n_2 I = n_0 + \left(\frac{n_2}{A_{eff}}\right) P \tag{12.2}$$

Here n_0 and n_2 are the linear and nonlinear refractive indexes, respectively. I and P are the optical intensity and optical power, respectively. The effective mode core area of the device is given by A_{eff}. For the microring and nanoring resonators, the effective mode core area varies from 0.10 μm² to 0.50 μm² [17].

When a Gaussian pulse is input and allowed to propagate within a fiber ring resonator, a resonant output is formed. Thus, the normalized output of the light field is the ratio between the output and input fields [$E_{out}(t)$ and $E_{in}(t)$] in each roundtrip, which can be expressed as [6, 7]

$$\left|\frac{E_{out}(t)}{E_{in}(t)}\right|^2 = (1-\gamma)\left[1 - \frac{(1-(1-\gamma)x^2)\kappa}{(1-x\sqrt{1-\gamma}\sqrt{1-\kappa})^2 + 4x\sqrt{1-\gamma}\sqrt{1-\kappa}\sin^2\left(\frac{\varphi}{2}\right)}\right]$$

(12.3)

Equation 12.3 describes a ring resonator with the ratio of the out power and input power equivalent to that of a Fabry–Perot cavity with an input and output mirror of field reflectivity $(1-\kappa)$ and a fully reflecting mirror. κ is the coupling coefficient, and $x = \exp(-\alpha L/2)$ represents a roundtrip loss coefficient. The linear and nonlinear phase shifts are $\phi_0 = kLn_0$ and $\varphi_{NL} = kL\left(\frac{n_2}{A_{eff}}\right)P$, respectively, where $k = 2\pi/\lambda$ is the wave propagation number in vacuum. L and α are the waveguide length and linear absorption coefficient, respectively. In this chapter, the iterative method is introduced to obtain results as shown in Eq. 12.3, which is similar to the condition when the output field is connected and input into other ring resonators.

The input optical field given by Eq. 12.1, i.e., a Gaussian pulse, is input into a nonlinear microring resonator. By using the appropriate parameters, a chaotic signal is obtained by using Eq. 12.3. To retrieve signals from the chaotic signals, or noise, we propose the use of an add/drop device with the appropriate parameters. The optical outputs of the add/drop filter are given by Eqs. 12.4 and 12.5.

$$\left|\frac{E_t}{E_{in}}\right|^2 = \frac{(1-\kappa_1)-2\sqrt{1-\kappa_1}\cdot\sqrt{1-\kappa_2}e^{-\frac{\alpha}{2}L}\cos(k_nL)+(1-\kappa_2)e^{-\alpha L}}{1+(1-\kappa_1)(1-\kappa_2)e^{-\alpha L}-2\sqrt{1-\kappa_1}\cdot\sqrt{1-\kappa_2}e^{-\frac{\alpha}{2}L}\cos(k_nL)}$$

(12.4)

$$\left|\frac{E_d}{E_{in}}\right|^2 = \frac{\kappa_1\kappa_2 e^{-\frac{\alpha}{2}L}}{1+(1-\kappa_1)(1-\kappa_2)e^{-\alpha L}-2\sqrt{1-\kappa_1}\cdot\sqrt{1-\kappa_2}e^{-\frac{\alpha}{2}L}\cos(k_nL)}$$

(12.5)

Here E_t and E_d represent the optical fields of the throughput and drop ports, respectively. The transmitted output can be controlled and obtained by choosing the suitable coupling ratio of the ring resonator, which is well derived and described by Yupapin and Suwancharoen [7]. Here $\beta = kn_{eff}$ represents the propagation constant, n_{eff} the effective refractive index of the waveguide, and

$L = 2\pi R$ the circumference of the ring, where R is the radius of the ring. In the following schematic diagram of a Gaussian soliton-generation system, new parameters will be used for simplification and where $\phi = \beta L$ is the phase constant. The chaotic signal cancellation can be managed by using specific parameters of the add/drop device, and the required signals in a specific wavelength band can be filtered and retrieved. κ_1 and κ_2 are the coupling coefficient of the add/drop filters, $k_n = 2\pi/\lambda$ the wave propagation number in vacuum, and $\alpha = 0.5$ dBmm^{-1} the waveguide (ring resonator) loss. The fractional coupler intensity loss is $\gamma = 0.1$. In case of an add/drop device, the nonlinear refractive index is neglected.

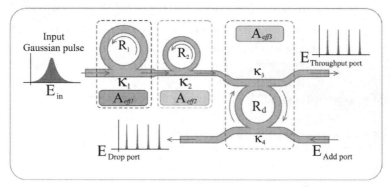

Figure 12.1   A schematic diagram of a Gaussian soliton–generation system for a laser gun and a laser tool. R_s: Ring radii, κ_s: Coupling coefficients, R_d: An add/drop ring radius, $A_{eff}s$: Effective areas

In operation, the light pulse is sliced into discrete signals that are amplified within the first ring, where more signal amplification can be obtained by using a smaller ring device (a second ring) as shown in Fig. 12.1. Thus, the required signals can be obtained via a drop port of the add/drop filter. An optical field in the form of Gaussian pulse from a laser source and a specified center wavelength is input into the system. From Fig. 12.2, the Gaussian pulse with center wavelength (λ_0) at 0.40 µm, pulse width (full width at half maximum, FWHM) of 20 ns, and peak power at 2 W is input into the system as shown in Fig. 12.2a. Large bandwidth signals can be seen within the first microring device as shown in Fig. 12.2b. Suitable ring parameters are used, such as ring radii $R_1 = 16.0$ µm, $R_2 = 5.0$ µm, and $R_d = 25.0$ µm. In order to make the system associate with the practical device [17], the parameters selected for the system are fixed at

n_0 = 3.34 (InGaAsP/InP), A_{eff} = 0.50 µm² and 0.25 µm² for the micro-ring and the add/drop ring resonator, respectively, α = 0.5 dBmm⁻¹, and γ = 0.1. In this investigation, the coupling coefficient (κ) of the microring resonator is varied from 0.55 to 0.90. The nonlinear refractive index of the microring used is n_2 = 2.2 × 10⁻¹⁷ m²/W. In this case, the attenuation constant used for the light propagated within the system (i.e., wave guided) is 0.5 dBmm⁻¹. After light is input into the system, the Gaussian pulse is sliced into a smaller signal, which spreads over the spectrum due to nonlinear effects [5] (Fig. 12.2a). A large bandwidth signal is generated within the first ring device. In applications, specific input or output wavelengths can be used and generated. For instance, the center wavelengths of an input pulse vary from 0.40 µm to 1.5 µm (Figs. 12.2–12.7).

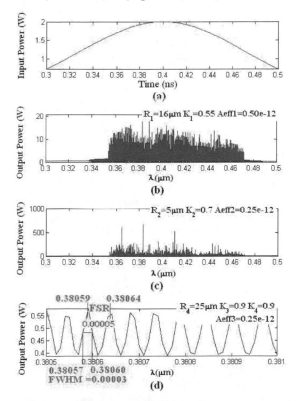

Figure 12.2   Results of the spatial pulses with center wavelength at 0.40 µm for (a) a Gaussian pulse, (b) large bandwidth signals, (c) large amplified signals, and (d) filtering and amplifying signals from the drop port.

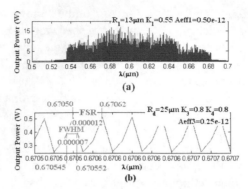

Figure 12.3   Results of the spatial pulses with center wavelength at 0.60 μm for (a) large bandwidth signals and (b) filtering and amplifying signals from the drop port.

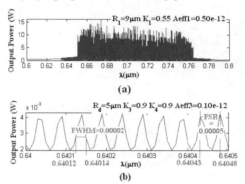

Figure 12.4   Results of the spatial pulses with center wavelength at 0.70 μm for (a) large bandwidth signals and (b) filtering and amplifying signals from the drop port.

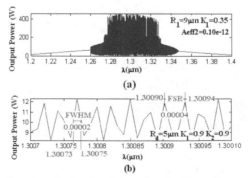

Figure 12.5   Results of the spatial pulses with center wavelength at 1.30 μm for (a) large bandwidth signals and (b) filtering and amplifying signals from the drop port.

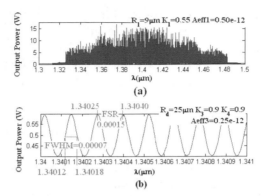

Figure 12.6   Results of the spatial pulses with center wavelength at
1.40 μm for (a) large bandwidth signals and (b) filtering and
amplifying signals from the drop port.

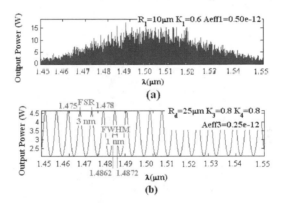

Figure 12.7   Results of the spatial pulses with center wavelength at
1.50 μm for (a) large bandwidth signals and (b) filtering and
amplifying signals from the drop port.

12.3   Laser Gun Mechanism

The results obtained show that a wide range of laser sources can be
generated along with the generation of a high-power light source. By
using the results obtained in Section 12.2, the laser tools system can
be designed as shown in Fig. 12.8. For safety reasons, the controlled
unit is important in this design. Therefore, each part of the system
has been carefully described in this section. After the input Gaussian
pulse is amplified and it reaches the specified value, the optical

energy gets stored in the ring resonator R_3 which has been designed and described in detail by Yupapin and Pornsuwancharoen [18]. The controlled parameters are the coupling coefficients κ_{31}, κ_3, and κ_{32}, which are chosen to control the output energy. The light sources generated and stored within the embedded system can be used in different applications.

For safety reasons, the coupling coefficients κ_{41} and κ_{42} are designed for low output energy. The laser energy can be increased by controlling the optical switch. The absorber is operated when the switch is in the open status mode so that it provides low output energy. The semi-automatic running mode can be arranged by keeping the optical switch in the operating position. Otherwise, it will be in manual operation status as shown in Fig. 12.8. Various applications, such as medical and kitchen tools, laser gun, and laser sword, can be operated by choosing suitable laser center wavelength, output energy, and beam size.

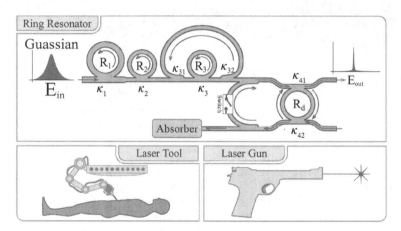

Figure 12.8 A diagram of the working mechanism of a laser tool. R_s: Ring radii, κ_s: Coupling coefficients, R_d: An add/drop ring radius, $A_{eff}s$: Effective areas.

12.4 Conclusion

In this chapter, we demonstrated how multi-wavelength laser sources can be generated by using a Gaussian pulse propagating within a microring resonator system and how pulses with center wavelengths ranging from 0.40 μm to 1.50 μm can be used in medical tools. By using a wider range of ring parameters, the spectral range of the output can be covered for a wider range instead of a fraction of a nanometer. A large increase in the peak power is seen when light propagates from a large effective core area to a small one. The output power ranges from a few watts to kilowatts and can be designed to meet the requirements of different applications. The advantage of the system proposed in this chapter is that a communication laser, specifically, a laser diode, can be pumped by a series of ring resonators as shown in Fig. 12.8 and the output power can be increased by using a nonlinear optical device. However, this is a simulation work in which a cooling part will be required to absorb the heat dissipated within the system.

References

1. J. Hecht, "Half a century of laser weapons," *Opt. Photonics News*, **20**(2), 14–21 (2009).

2. D. Deng, and Q. Guo, "Ince-Gaussian solitons in strongly nonlocal nonlinear media," *Opt. Lett.*, **32**, 3206–3208 (2007).

3. G. Xia, Z. Wu, and J. Wu, "Effect of fiber chromatic dispersion on incident super-Gaussian pulse transmission in single-mode fibers," *Chin. J. Phys.*, **41**(2), 118–120 (2003).

4. S. Supparpola, Y. Sun, and S. A. Chiramida, "Gaussian pulse decomposition: An intuitive model of electrocardiogram waveforms," *Ann. Biomed. Eng.*, 252–260 (1997).

5. P. K. A. Wai, and K. Nakkeeran, "On the uniqueness of Gaussian ansatz parameters equations: Generalized projection operator method," *Phys. Lett. A*, **332**, 239–243 (2004).

6. P. P. Yupapin, P. Saeung, and C. Li, "Characteristics of complementary ring-resonator add/drop filters modeling by using graphical approach," *Opt. Commun.*, **272**, 81–86 (2007).

7. P. P. Yupapin, and W. Suwancharoen, "Chaotic signal generation and cancellation using a microring resonator incorporating an optical add/drop multiplexer," *Opt. Commun.*, **280**, 343–350 (2007).

8. P. P. Yupapin, N. Pornsuwancharoen, and S. Chaiyasoonthorn, "Attosecond pulse generation using nonlinear microring resonators," *Microw. Opt. Technol. Lett.*, **50**, 3108–3111 (2008).

9. N. Pornsuwancharoen, and P. P. Yupapin, "Generalized fast, slow, stop, and store light optically within a nanoring resonator," *Microw. Opt. Technol. Lett.*, **51**, 899–902 (2009).

10. Y. S. Kivshar, and B. Luther-Davies, "Dark optical solitons: Physics and applications," *Phys. Rep.*, **298**, 81–197 (1998).

11. A. Hasegawa (ed), "Massive WDM and TDM soliton transmission systems," Kluwer Academic Publishers, Boston (2000).

12. G. P. Agrawal, "Nonlinear fiber optics," 4th edition, Academic Press, New York (2007).

13. Y. A. Simonov, and J. A. Tjon, "Soliton-soliton interaction in confining models," *Phys. Lett. B*, **85**, 380–384 (1979).

14. J. K. Drohm, L. P. Kok, Y. A. Simonov, J. A. Tjon, and A. I. Veselov, "Collision and rotation of solitons in three space-time dimensions," *Phys. Lett. B*, **101**, 204–208 (1981).

15. T. Iizuka, and Y. S. Kivshar, "Optical gap solitons in nonresonant quadratic media," *Phys. Rev. E*, **59**, 7148–7151 (1999).

16 .R. Ganapathy, K. Porsezian, A. Hasegawa, and V. N. Serkin, "Soliton interaction under soliton dispersion management," *IEEE J. Quant. Electron.*, **44**, 383–390 (2008).

17. Q. Xu, and M. Lipson, "All-optical logic based on silicon micro-ring resonators," *Opt. Express*, **15**(3), 924–929 (2007).

18. P. P. Yupapin, and N. Pornsuwancharoen, "Proposed nonlinear micro-ring resonator arrangement for stopping and storing light," *Photon. Technol. Lett.*, **21**(6), 404–406 (2009).

Index

AD *see* add/drop filter
add/drop device 53, 55, 88, 89, 100, 112, 155
add/drop filter (AD) 21–23, 26, 27, 29, 30, 32, 37, 55, 89, 100, 146, 155, 156
add/drop filter radius 100
add/drop multiplexer 83, 86, 97, 124
AFM *see* atomic force microscopy
all-optical adder/subtractor 35, 36, 38, 40, 42, 44, 46
all-optical circuit 35, 36
all-optical logic gates 139, 146–149
all-optical logic XOR/XNOR gates 139, 140, 142, 144, 146, 148
amplified tweezers 91, 92
artificial kidney 123, 128
atomic force microscopy (AFM) 4, 5
atom trapping 115, 117
attenuation coefficient 23–25, 53, 54, 65, 112, 113

band gap 7–9
  photonic 9
BC *see* beam combiner
beam combiner (BC) 40
beam splitter (BS) 40
blood cells 128
blood cleaner 128
blood cleaner on-chip design 123, 124, 126, 128, 130, 132
blood concentrations 127, 132

blood flows 127, 128
blood wastes 125, 128, 132
bright soliton pulses 27, 29, 36, 40, 75, 79
BS *see* beam splitter
buffer 98, 101, 103, 104
  optical 97

center wavelengths 27, 56, 64, 102, 116, 129, 132, 154, 156–159
  laser 160
  specified 156
center wavelengths of tweezers 93, 102
chaotic signals 21, 155
chronic kidney disease (CKD) 123
ciphertext 29, 31
  optical 30
circuit
  all-optical 35, 36
  electronic 35, 36, 43
circulated light fields 24, 25, 54, 112, 113
CKD *see* chronic kidney disease
clusters 7, 8
coefficients
  field coupling 37, 140
  intensity coupling 53, 54
computational domain 12, 13, 15–17
  2-D 12, 13
control fields 38, 141
control port 22, 29, 55, 56, 87, 110, 115, 127, 132, 142
control port signals 93, 99, 102, 126